4-H Robotics: Engineering for Today and Tomorrow

Junk Drawer Robotics

Level 3

Mechatronics

Presenter's Activity Guide

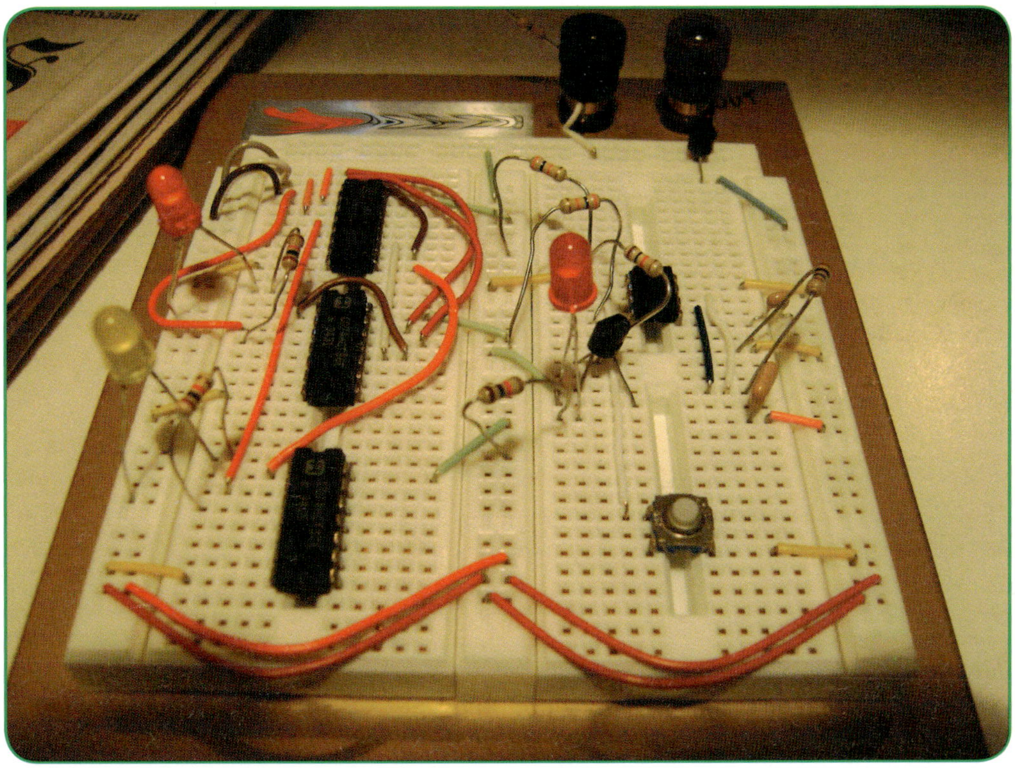

Junk Drawer Robotics Level 3
Mechatronics

Table of Contents

Introduction to Junk Drawer Robotics

The goal of *Junk Drawer Robotics* is to make science, engineering, and technology engaging and meaningful in the lives of young people. The activities do this by encouraging youth to use the processes and approaches of *science*; the planning and conceptual design of *engineering*; and the application of *technology* in their personal portfolios of skills and abilities.

The Junk Drawer Robotics curriculum is divided into three levels or books, each around a central theme related to robotics design, use, construction, and control. Each level starts out with background information on working with youth, curriculum elements, and a focus on the concepts to be addressed.

- In Book 1, the theme is robotic arms, hands, and grippers.
- In Book 2, the theme is moving, power transfer, and locomotion.
- In Book 3, the theme is the connection between mechanical and electronic elements.

Modules within each book target major concepts of the theme. The modules each contain five or more activities that help the members develop an understanding of the concepts, create solutions to challenges, and develop skills in constructing alternatives.

Your role as a presenter in this curriculum is different from the normal role of the teacher that we may know from a school setting. It is not about mere transfer of knowledge from teacher to student. It is about assisting learners in developing their own knowledge and problem solving skills. This is done by bringing together a scientific inquiry and engineering design approach to learning. Youth will learn about a topic by exploration. When given a problem, they will design a solution. Then, using what they have learned and designed, they will build or construct a working model.

The presenters of this curriculum may be teachers, afterschool staff, or volunteers, including teens working with younger youth. In the case of teens, an adult coach or mentor could provide support, training, and guidance to teams of two or three teens who present the activities together. As a presenter, you will assist the youth in understanding the processes of science, engineering, and technology by how you ask questions and have them share their ideas, designs, and results.

The robotics curriculum is designed around three themes of science, engineering, and technology. Each module has activities in each of these three areas. As a framework, *4-H Junk Drawer Robotics* uses these simple definitions developed by Anne Mahacek, former 4-H teen member who is now a mechanical engineer and grad student in mechatronics.

To Learn: **Science** is finding out how things work.

To Do: **Engineering** is using what you found out to design something to work.

To Make: **Technology** is using tools and processes to make something work.

What Is 4-H Science?

The National 4-H Science initiative addresses America's critical need for more scientists and engineers by engaging youth in activities and projects that combine nonformal education with design challenges and hands-on, inquiry-based learning in a positive youth development setting. These experiences engage youth and help them build knowledge, skills, and abilities in science, engineering, mathematics, and technology.

All 4-H Robotics activities and projects are:

- Based on National Science Education Standards and Standards for Technological Literacy,
- Focused on developing abilities in science, mathematics, engineering, and technology,
- Led using the Experiential Learning Model,
- Tied to developing Life Skills for youth, and
- Delivered in a positive youth development context by trained and caring adults.

National Science Education Standards (NSES)

The National Science Education Standards present a vision of a scientifically literate person and present criteria for our education system that will allow that vision to become reality (National Research Council, 1996). NSES outline what students should know, understand, and be able to do as they progress through their science education. Emphasis has shifted from being solely on "the content to be learned" to include "**how** students learn" and "**how** the content is presented."

Scientific Inquiry refers to the diverse ways in which scientists study the natural world and propose explanations of the world based on the evidence derived from their work (National Research Council, 1996). But scientific inquiry is not limited to the work of scientists. Young people driven by curiosity and given a structure can pose questions, make observations, analyze data, and offer their own explanations. Supporting youth in developing the skills and understanding necessary to engage in scientific inquiry is a central focus of the NSES and 4-H Science.

Standards for Technological Literacy (STL)

Standards for Technological Literacy define what youth should know and be able to do in order to be technologically literate (International Technology Education Association, 2000). Technological literacy is important to all youth whatever path they pursue in life. They offer a common set of expectations for what students should learn in the study of technology and what is developmentally appropriate at different ages. The 20 standards address five areas: the nature of technology, technology in society, design, abilities for a technological world, and understanding the designed world. The activities in *Junk Drawer Robotics* help students develop literacy in each of these areas.

Science, Engineering and Technology Abilities

According to Horton, Gogolski, and Warkentien (2007), effective teaching in science, engineering, and technology must focus on how youth learn the content **and** how the material is taught. Based on a review of the science, engineering, and technology education literature, the authors set forth 30 important processes and refer to them as Science, Engineering and Technology Abilities (Horton, Gogolski, & Warkentien, 2007). This set of science, engineering, and technology life skill outcomes is emphasized throughout the 4-H Robotics curriculum. Examples of some of the 30 abilities that are developed in this curriculum include: observe, categorize, organize, infer, question, predict, evaluate, use tools, measure, test, redesign, collaborate, summarize, and compare. Each module in *Junk Drawer Robotics* identifies particular abilities in science, mathematics, engineering, and technology that are focused on in that module.

Experiential Learning Process

Experiential learning allows young people to create or develop their own answer to a question instead of repeating "the answers" (Maxa, et. al., 2003). In the experiential learning process, youth are encouraged to think, explore, question, and develop decisions. Two important components of the experiential learning process are a period of reflection, during which the learner shares and processes the experience, and the application of new learning in "real world" situations. The experiential learning model contains five steps: experience (doing), share (what happened?), process (what's important), generalize (so what?), and apply (now what?) (Maxa, et. al., 2003).

The Experiential Model

This model helps leaders formulate activities to reflect the DO, REFLECT, APPLY in five steps.

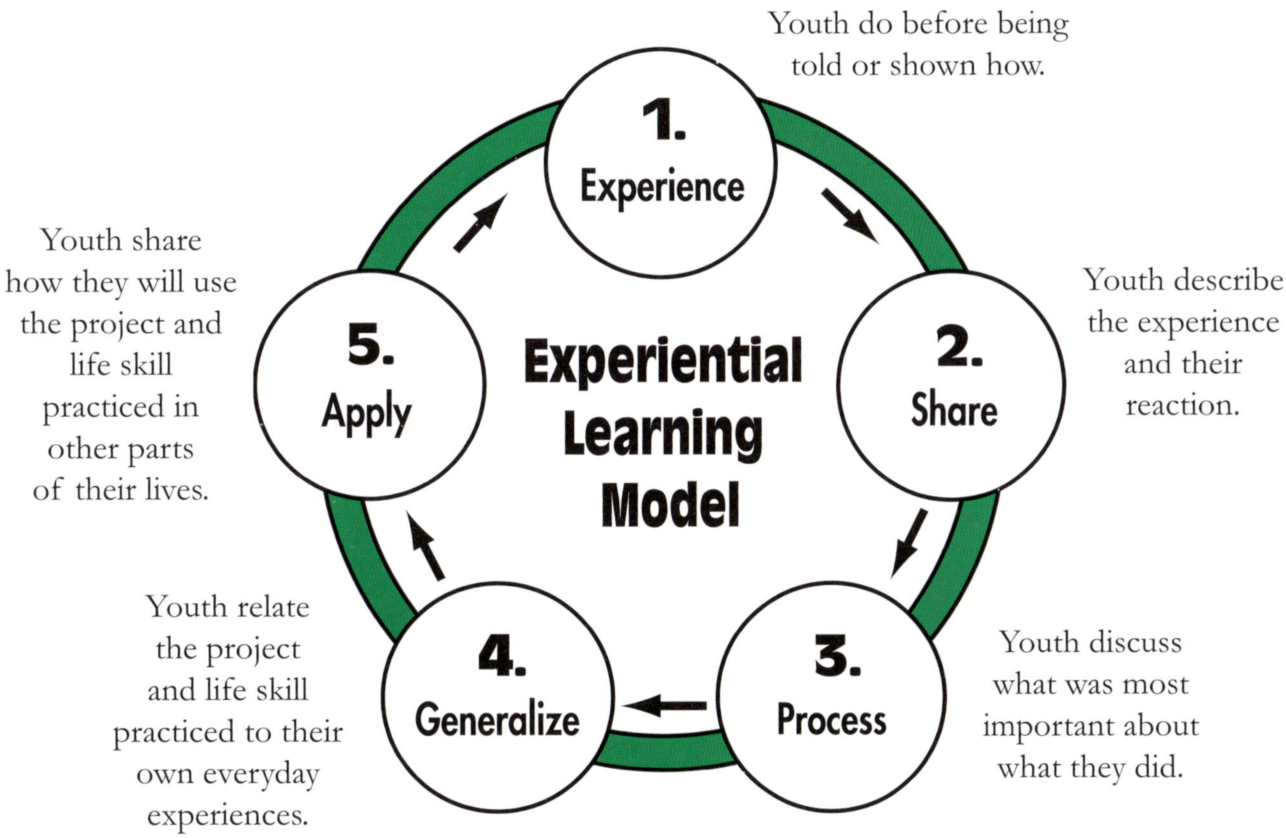

Pfeiffer, J. W., & Jones, J. E., *Reference Guide to Handbooks and Annuals* 1983, John Wiley & Sons, Inc. Reprinted with permission from John Wiley and Sons, Inc.

Facilitation of the Experiential Learning Process

The key to the experiential learning process is that youth seek answers to questions rather than being given answers. This process requires facilitation instead of instruction. The role of adults in the experiential learning process is to facilitate the learning process, which means they become co-learners (Maxa, et. al., 2003).

Questioning Strategies

Questions suggested by the facilitator are designed to help promote discussion and engagement with life skills and science, engineering, and technology abilities. The goal is to have the questions reside with the learner. Questions should promote discussion, interaction, and stimulate learner thinking. They should encourage ideas, speculation, and the formation of hypotheses. This type of questioning will not lead to a single right answer, but it will promote deeper understanding. It is important to allow adequate time for questions and discussions that engage youth and enhance learning. The process cannot be rushed.

Positive Youth Development

High quality 4-H Science programming provides valuable benefits in engaging youth. These programs also give youth the opportunity to engage in a positive youth development setting. Positive youth development occurs from an intentional process that promotes positive outcomes for young people by providing opportunities, choices, caring relationships, and the support necessary for youth to fully participate in families and communities. Simply, positive youth development seeks to develop young people as resources instead of problems to be managed (Lerner, 2005). Creating a positive youth development setting requires that youth are able to develop a sense of competence in social, academic, cognitive, health, and vocational aspects of life, a feeling of self-efficacy, a positive bond with a caring adult, a respect for societal and cultural norms, a sense of sympathy and empathy for others, and make contributions to self, family, communities, or society (Lerner, 2005).

4-H National Headquarters has identified four essential elements of positive youth development. They include:

- Belonging – To know they are cared about by others
- Mastery – To feel and believe they are capable and successful
- Independence – To know they are able to influence people and events
- Generosity – To practice helping others through their own generosity

To ensure that youth are engaged in a positive youth development setting, it is critical that learning includes the 4-H essential elements of positive youth development. This will give youth the opportunity to develop positive youth/adult relationships, practice life skills, and engage in the experiential learning model, which can promote mastery, independence, and generosity.

4-H Life Skills

Life skills are important in helping youth become self-directing, productive, contributing members of society (Maxa, et. al., 2003). 4-H programming strives to give youth developmentally appropriate opportunities to experience life skills, to practice them until they are mastered, and use these skills throughout a lifetime (Hendricks, 1998). The Targeting Life Skills model is based on the 4-H clover, which represents head, heart, hands, and health.

Targeting Life Skills Model

The "Targeting Life Skills Model" helps identify developmental life skills in 4-H and Youth Development Educational programs. They are grouped by:

Head Managing, Thinking
Heart Relating, Caring
Hands Giving, Working
Health Living, Being

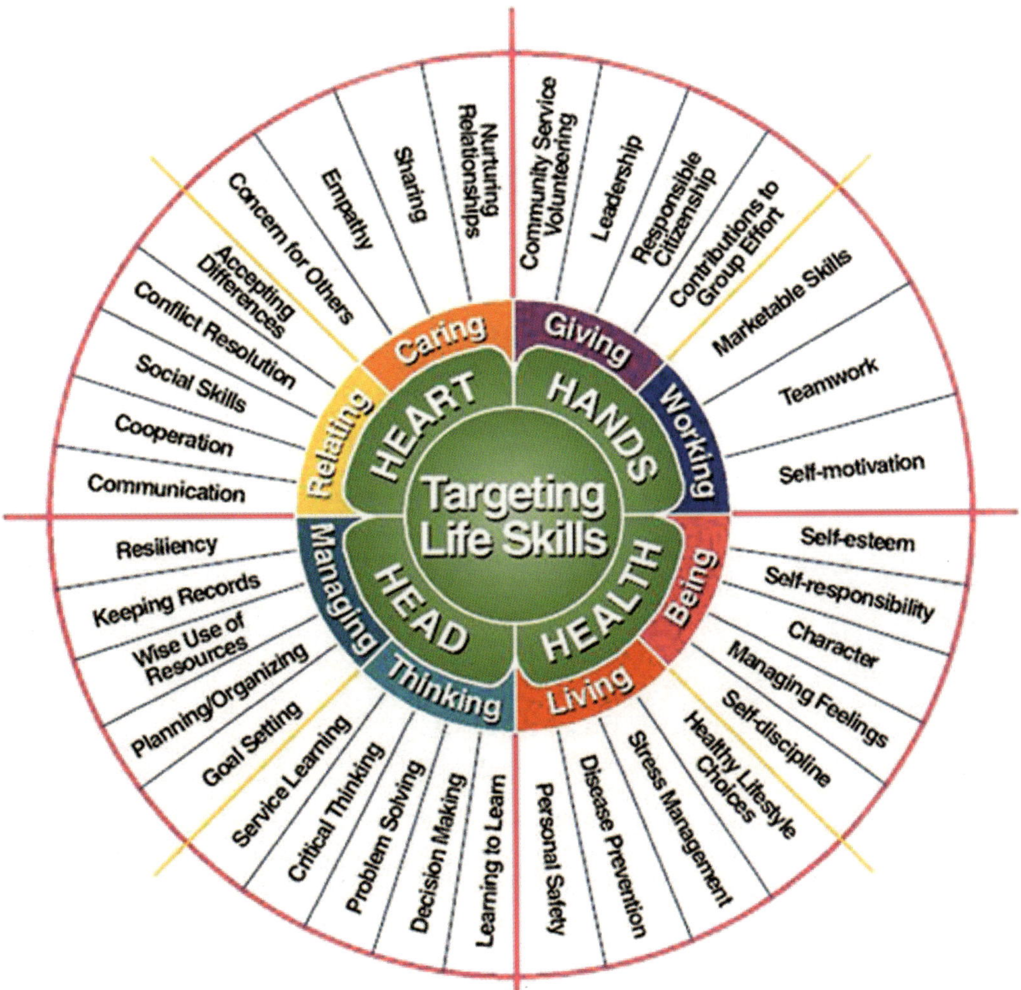

Graphic provided by Iowa State University Extension

What You Will Need for Junk Drawer Robotics

1. Trunk of Junk

The Trunk of Junk is a collection of items that will be used throughout this project. You may collect many of the items before you begin or add items as needed. Items you might collect are:

- Cardboard tubes from gift wrap, aluminum foil, paper towels, etc.
- Stationery supplies, like paper clips, binder clips, paper brads, etc.
- Clothespins, used household items, old toys, and items to take apart
- Coffee stirrers (sticks and straws), drinking straws, paper and plastic cups
- Construction paper, copy paper, graph paper, etc.
- Various pieces of cardboard; flats, boxes, tubes
- Assortment of small bolts, nuts, washers, and screws
- Wooden sticks (paint sticks, plant stakes, craft sticks, etc.) **Note**: Wooden sticks with holes are very useful in building. Holes can be predrilled or youth can drill holes as needed if you have appropriate tools. The website at *www.4-H.org/curriculum/robotics* includes tips on drilling holes in craft sticks.

2. Activity Supplies

Each activity has a list of the items to be prepared for that particular activity. Advance preparation or photocopying may be required. Some of the supplies may come from the Trunk of Junk, and some will be specific to the activities, but most will need to be organized before the meeting. Specific instructions are included in each activity.

3. Toolbox

This is a selection of basic hand tools that can be used throughout the project. These may be stored together in a toolbox, or collected before each meeting as needed. Commonly used tools include:

- Glue, tape, scissors (one or two per group)
- Low-temperature glue gun (two or three to share)
- Hand drill with different size bits
- Small hacksaw (to cut dowels and small boards)
- Pliers, wire cutters, screwdrivers, small hammer
- Bench hook or work surface to protect tables and chairs
- Paper towels, brush, cleanup supplies
- Punches (leather, craft, hole)

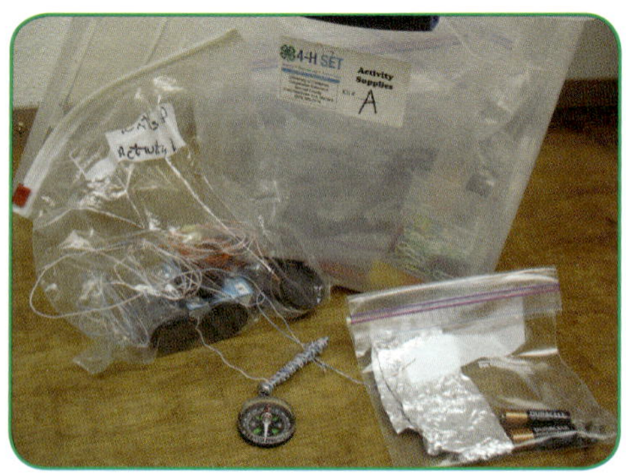

4. The Presenter's Activity Guide

This booklet is one of the three Presenter's Activity Guides: Give Robots a Hand, Robots on the Move, and Mechatronics. Each guide includes three to five modules made up of a series of activities. Three types of activities are within each module. In the To Learn activities, the focus is on gaining knowledge as a scientist. The To Do activities promote engineering design and innovation. And the To Make activities develop technological skills.

5. Robotics Notebook

There is one Robotics Notebook for the three levels of the *Junk Drawer Robotics* curriculum. Using the notebook encourages members to think like scientists and engineers. In their notebook, they will record their ideas, collect data, draw designs, and reflect on their experiences. It also provides specific information for the challenges. Each youth should have his or her own Robotics Notebook. If this is not possible, a blank notebook can be substituted. Graph paper will work best. Youth will have to record both the questions and their responses in a blank notebook, and leaders will have to make copies of supporting material.

6. Website and On-line Resources

The 4-H Robotics website, *www.4-H.org/curriculum/robotics*, provides supporting information, including background about robotics concepts, tips, and resources for leaders. You will find detailed tips on gathering tools, locating supplies, and instructions on how to make some of your own materials for *Junk Drawer Robotics*. There are also resources specifically for coaches or mentors working with a team of teens as they present the *Junk Drawer Robotics* modules. The role of a coach or mentor is to provide the teens with back-up support and resources. The website also has an overview of how the *Junk Drawer Robotics* activities are related to other parts of *4-H Robotics: Engineering for Today and Tomorrow* curriculum and suggestions for implementing the curricula in different settings like camps, events, clubs, or afterschool programs.

7. Additional 4-H Robotics Opportunities

Extend your experience with robotics with another part of the *4-H Robotics: Engineering for Today and Tomorrow* curriculum. *4-H Virtual Robotics* provides the opportunity to utilize an interactive computer game environment to learn about the science and engineering of robots. *4-H Robotics Platforms* provides a curriculum to use as you explore a robotics kit and learn to design, build, and program a robot. For more information or to order the curriculum, visit the website at *www.4-H.org/curriculum/robotics*.

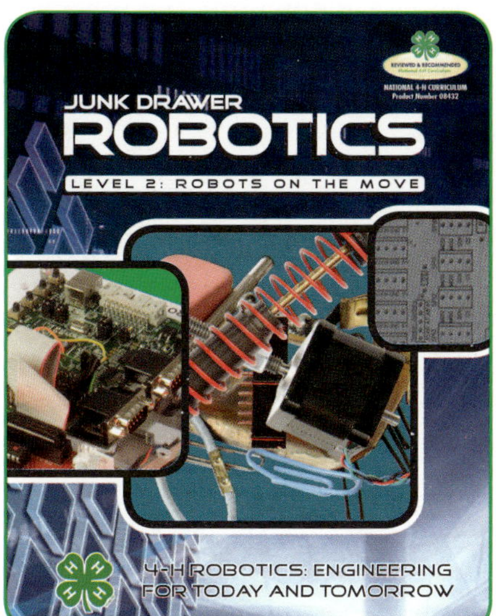

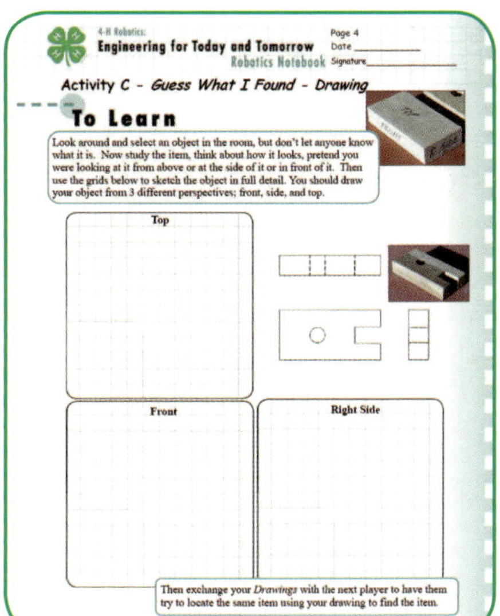

Focus for Mechatronics

Introduction to Level 3 Junk Drawer Robotics: Mechatronics

The combination of mechanical and electrical systems helps to provide the control and automation we come to expect in a robot. The term Mechatronics implies a synergy of bringing together many systems from computers and sensors to mechanics and power sources.

In this level, Mechatronics, we will explore how we can build on our learning and experiences from designing and building robotic arms and rovers in Levels 1 and 2. We will begin to understand sensors, test simple electrical circuits, and learn about analog and digital systems through activities in binary and logic systems. We will introduce simple electronic components and build basic test circuits to see how the components work. We will investigate basic elements of programming and instructions for robotic computer control.

The final module in Mechatronics is an open challenge for the members based on combining all three levels. We hope that they use the knowledge they have gained during the To Learn activities, that they apply the problem solving strategies they uncovered in the To Do design activities, and that they rely on using the skills in technology from the To Make activities to create a robotics application of their own.

Big Ideas

Module 1

An electronic circuit is a closed path through which an electric charge can flow. It typically consists of a power source (battery), conductors (wire), a load (bulb), and a control (on/off switch).

The interaction between the "sense," "think," and "act" components of what a robot does can be accomplished using electrical signals.

Module 2

Simple sensors can help control robot behavior.

Many different types of sensors are available to provide information to the robot.

Sensors can enable robots to work autonomously (without human control).

Module 3

While humans typically use a base 10 number system, digital systems with a base 2 system (on/off) are more efficient for computers.

Logical operators, used in Boolean logic, form the basis of all modern digital electronics.

Basic electronic circuit design, including components of batteries, wires, conductors, resistors, lights, capacitors, diodes, transistors, and switches.

Module 4

Programming allows us to control robot behavior.

Basic logic elements in programming allow us to predict outcomes.

Flowcharts are often used in programming to help design and clarify instructions.

Input, output, if-then-else, loops, goto, and other commands are core components of simple programs.

Module 5

Robots can sense, plan (think), and act. Generally, they are designed to perform a specific job better than it could be done by a person or another machine.

How a robot behaves or acts is a product of its design. The robot's structure, components, and programming determine its function.

Engineering design is a purposeful process of generating and evaluating ideas that seek to develop and implement the best possible solution to a given problem.

Connected Ideas

Electronic Systems/Circuits is presented in

Virtual Robots: Electronics: Power Up!,
Junk Drawer Robotics: Watt's Up?
Junk Drawer Robotics: Circuit Training, and
Junk Drawer Robotics: It's Logical.

Sensor/Sensing is presented in

Virtual Robotics: Robot Behavior,
Virtual Robotics: Sensor Exploration,
Junk Drawer Robotics: Come to Your Senses,
Robotic Platforms: Sensor Games, and
Robotic Platforms: Simple Sensor Robot.

Programming/Instruction is presented in

Virtual Robotics: Robot to the Rescue,
Junk Drawer Robotics: Come to Your Senses,
Junk Drawer Robotics: Do What I Say, and
Robotic Platforms: Move the Robot.

National Science Education Standards (NSES)

Unifying Concepts and Processes:

Evidence, models, and explanation
Constancy, change, and measurement
Systems, order, and organization

Science as Inquiry:

Abilities necessary to do scientific inquiry

Physical Science Standards:

Light, heat, electricity, and magnetism

Science and Technology:

Abilities of technological design
Understanding about science and technology

Standards for Technological Literacy (STL)

The Nature of Technology

Core concepts of technology
Relationships and connections

Technology and Society

Social effects of technology
Development and use of technology

Design

Attributes of design
Engineering design
Problem solving

Abilities for a Technological World

Apply the design process

The Designed World

Energy and power technologies
Information and communication technologies

SET Abilities Developed in this Level

- Build/Construct
- Compare/Contrast
- Communicate/Demonstrate
- Decision Making
- Design Solutions
- Draw/Design
- Invent/Implement Solutions
- Observation
- Sequencing
- State a problem
- Redesign

Life Skills Practiced in this Level

- Communication
- Contribution to Group Effort
- Cooperation
- Critical Thinking
- Keeping Records
- Planning/Organizing
- Problem Solving
- Sharing
- Self-Discipline
- Teamwork

Module 1: Circuit Training

Overview of Activities in this Module

To Learn

Activity A – Series/Parallel
Activity B – Off and On
Activity C – Direction of Flow

To Do

Activity D – Forward and Reverse Design Team

To Make

Activity E – Forward and Reverse Build Team

Note to Leader

Circuit — In a circuit, an electrical flow or path is needed from the + (plus) battery contact through the item (bulb, motor, light, hair dryer, etc.) and back to the – (negative) contact of the battery. When there is a complete path, the electricity will flow through the circuit and the light will glow, the motor will turn, and the hairdryer will get hot.

Switches — Switches are used in a circuit to control the flow by opening or breaking the flow, which turns off the item. When the light switch is flipped off, the circuit is opened. No electricity will flow, and the light goes off. This switch allows control over the electrical flow and the light.

Series circuit — In a series circuit, one wire (conductor) goes from the power source (battery, transformer, etc.) to the item and back, as in Example A. In this example, the two lights have one conductor going into and out of each light (one loop). This construction saves wire, but if one light burns out, the circuit will open and the other light will go out.

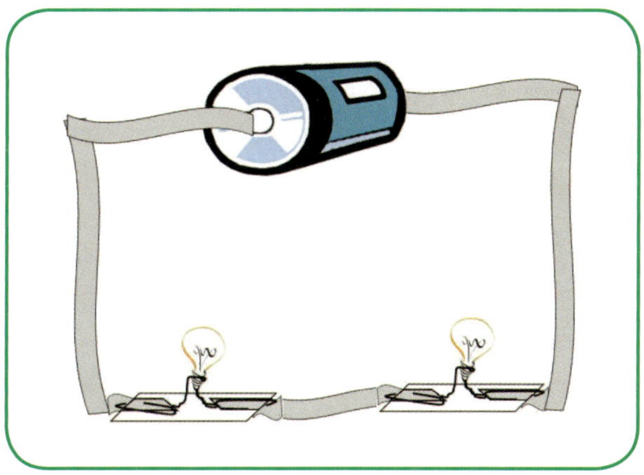

Example A

Parallel circuits — In a parallel circuit, two wires (conductors) go to each item, as in Example B. In this example, each of the two light bulbs has an individual path (multiple loops) for the flow of electricity. Even if one light burns out, the other retains its electrical flow and remains lit.

There are combinations of series/parallel circuits for multiple items. Multiple **batteries** also can be connected in either a series or parallel circuit. Batteries connected in series will add voltages. However, if a group of batteries is connected together in a series and create too high a voltage, the light bulbs will burn out.

As stated before, an electrical switch is used to control the on/off flow of electricity in a circuit. The switch can either be on (flow) or off (no flow). The opening and closing of the switch can control the lights, motors, or other electrical items in the circuit. Electronic components like transistors also are used for switching and control of lights, motors, and other circuits. By using electronic circuits, a whole array of options for programming, sensor data, and control can be automated for the robot.

In this module, we will begin to look at simple electrical circuits and physical switches to control them. A simple switch can be made from a few metal parts (conductors) and some insulating parts (wood, plastic, air). A knife switch is one example with contact A, contact B, and a lever C, as shown. The contacts A and B are wired into the circuit. The lever C is moved to touch both contacts to allow the electricity to flow, and pulled up to stop the flow. In this module you will assist the participants in exploring series and parallel circuits, switches, and in designing and creating their own switches using craft sticks, aluminum foil, paper brads, and paper clips.

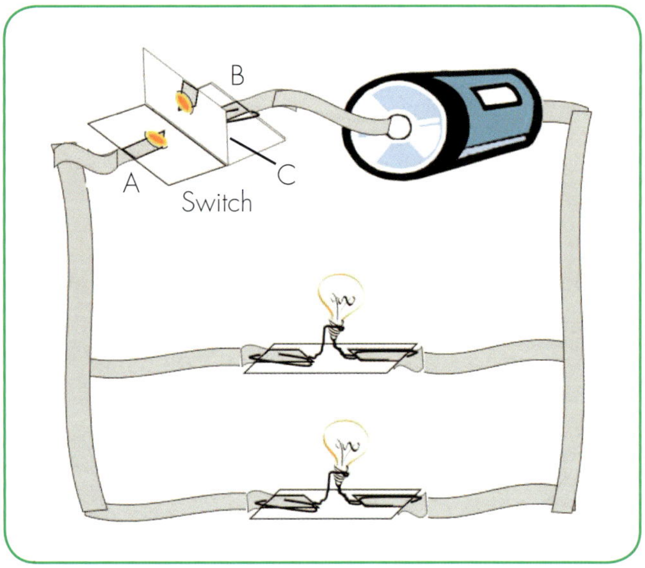

Example B

What you will need for Module 1: Circuit Training

- Robotics Notebook for each youth
- Trunk of Junk, see page 8
- Tool Box
- Activity Supplies
  - Copies of handout/poster of electrical circuits on page 18
  - Copies of handout/poster of electrical switches on page 20
  - Wire or aluminum foil and craft sticks with holes
  - Paper brads, paper clips, rubber bands
  - Masking, clear, or electrical tape
  - Holiday mini and holiday LED lights
  - 1.5-volt batteries – AA, C, or D size
  - Toy electric motor (1.5 volt to 6 volt will work)
  - A wheel, gear, propeller, or cardboard disk to fit on the toy motor
  - Backing or base for holding circuit elements being designed; can use pegboard, cardboard, or plastic canvas for this base
  - Drill and bits

Note: See See the 4-H Robotics website at *www.4-H.org/curriculum/robotics* for more information about electrical circuits and suggestions for making foil connections using craft sticks, lamp holders, switches, and battery holders.

Timeline for Module 1: Circuit Training

Activity A – Series/Parallel

- Activity will take approximately 20 minutes.
- Divide youth into groups of two or three.

Activity B – Off and On

- Activity will take approximately 20 minutes.
- Divide youth into groups of two or three.

Activity C – Direction of Flow

- Activity will take approximately 20 minutes.
- Divide youth into groups of three or four.

Activity D – Forward and Reverse Design Team

- Activity will take approximately 20 minutes.
- Divide youth into groups of two or three.

Activity E – Forward and Reverse Build Team

- Activity will take approximately 40 minutes.
- Use the same groups from Activity D, Forward and Reverse Design Team.

Performance Tasks For Youth

You will learn about the differences between series and parallel electric circuits.

You will create an on/off switch for a basic circuit.

You will experiment with circuits and control the direction of electrical flow.

You will plan and design a double pole double throw (DPDT) switch to change the circuit's polarity.

You will use your DPDT design and build a working switch that you can attach to a simple circuit.

Focus for Module 1: Circuit Training

Big Ideas

- An electrical circuit is a closed path through which an electric charge can flow. It typically consists of a power source (battery), conductors (wire), a load (bulb), and a control (on/off switch).
- The interaction between the "sense," "think," and "act" components of what a robot does is accomplished using electric signals.

NSE Standards

- Constancy, change, and measurement
- Light, heat, electricity, and magnetism
- Abilities of technological design
- Understanding about science and technology

STL

- Relationships and connections
- Engineering design
- Energy and power technologies

SET Abilities

- Build/construct
- Compare/contrast
- Draw/design

Life Skills

- Communication
- Critical thinking
- Keeping records
- Planning/organizing

Success Indicators

Youth will build simple operating electric circuits using parallel and series designs.

Youth will be able to turn off and on a circuit they created to power a light and/or motor.

Youth will change the direction of electrical flow in a circuit.

Youth will plan, design, and share what they've learned about a double pole double throw switch.

Youth will create a DPDT switch that will reverse a motor's rotation.

Activity A – Series/Parallel

Performance Task For Youth

You will learn about the differences between series and parallel electric circuits.

Success Indicators

Youth will build simple operating electric circuits using parallel and series designs.

List of Materials Needed

- Robotics Notebook
- Activity Supplies
  - Craft sticks, brass brads, wire or aluminum foil, paper clips, tape, rubber bands, three holiday mini lights, and batteries
  - A base to attach circuit elements made of pegboard, cardboard, or plastic canvas

    Note: See the 4-H Robotics website at *www.4-H.org/curriculum/robotics* for more information about electrical circuits and suggestions for making foil connections using craft sticks, lamp holders, switches, and battery holders.
  - Copies of handout/poster of electrical circuits on page 18

Activity Timeline and Getting Ready

- Activity will take approximately 20 minutes.
- Divide youth into teams of two or three.
- Prepare two identical packets per team, including a base or backing material about 6 inches square, three small holiday mini lights, two craft sticks (with holes), and one battery.
- Provide other materials, such as brads, paper clips, aluminum foil, or rubber bands as needed, or allow teams to select materials from the supply area.
- Share handout or refer to the Robotics Notebook for information on types of circuits. A series circuit consists of a one-loop circuit with each item connected to the next, providing one path for the electrical flow. A parallel circuit consists of a multiloop circuit in which each item can make its own path (loop) for electrical flow.

Experiencing

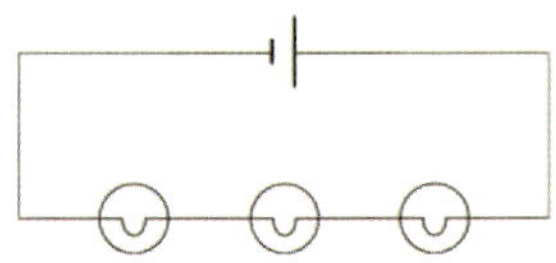

1. Ask the youth to build a three light series circuit using one of the prepared packets. For this exercise, none of the bulb wires should be directly connected to another, and the whole circuit should be attached to a cardboard or plastic canvas. This will require the use of two aluminum foil-covered craft sticks, paper brads, paper clips, etc., to make the connections. When finished, test the circuit with the battery. Save this circuit setup for later use.

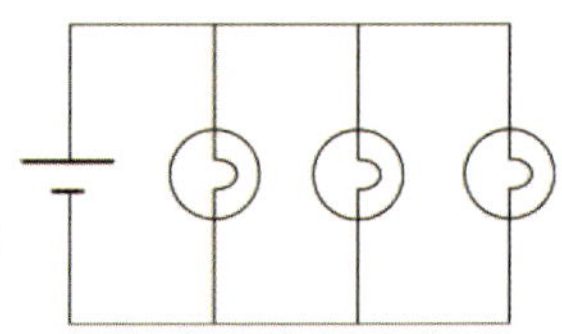

2. Ask youth to build a second three light circuit with the second prepared packet, but this time as a parallel circuit. Once again, none of the bulb wires should be directly connected to another, and they should be attached to a cardboard or plastic canvas base. This will require the use of two aluminum foil-covered craft sticks, paper brads, paper clips, etc., to make the connections. When finished, test the circuit with the battery. Save this circuit setup for later use.
3. Ask each team to compare the designs of their two circuits with the other groups' creations — how they are alike and how they are different. Students should record their observations in the Robotics Notebook.
4. Ask participants to modify one of their circuits to create a hybrid with both series and parallel sections in the same circuit.

Sharing and Processing

As the facilitator, help guide youth as they question, share, and compare their observations. Have participants record their experience in their Robotics Notebook, including notes about new discoveries and knowledge. If necessary, use more targeted questions to prompt particular points.

- How are the lights affected by circuit type? Brightness/dimness?
- When one light goes out, what happens to the others? (Note: When one light goes out in a series circuit, all the lights go out. In a parallel circuit, even if one light goes out, the others stay on.)
- Which circuit do you think would be better for building a robot? Why?

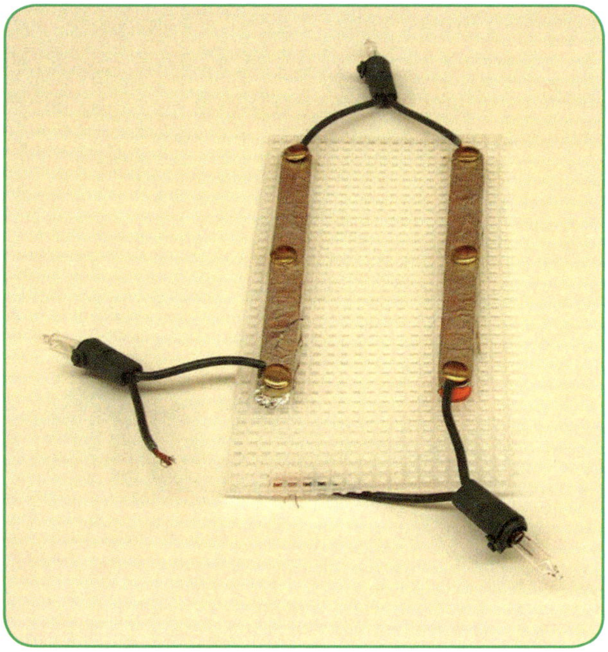

Generalizing and Applying

- Where have you seen or used series circuits? (Series: In some older holiday decoration lights, when one bulb goes out, they all go out.)
- When have you seen or used parallel circuits? (Parallel: In stadium lights, when some burn out, the rest stay lit.)
- Share other types of circuits you've seen. Are they similar or different to the one you've created?
- Youth can apply what they have learned in Activity B.

Note: Have the participants save these two bases and circuits for use in Activity B, Off and On, and Activity C, Direction of Flow.

Electrical Circuits

There are many ways to create an electrical circuit for the flow of electricity (electrons) to do things like buzz a buzzer, make a motor turn, or light a lamp bulb.

A simple electrical circuit can be made from a battery, wires (connectors), and a light bulb (lamp).

This circuit can be drawn using symbols: These drawings are called schematics.

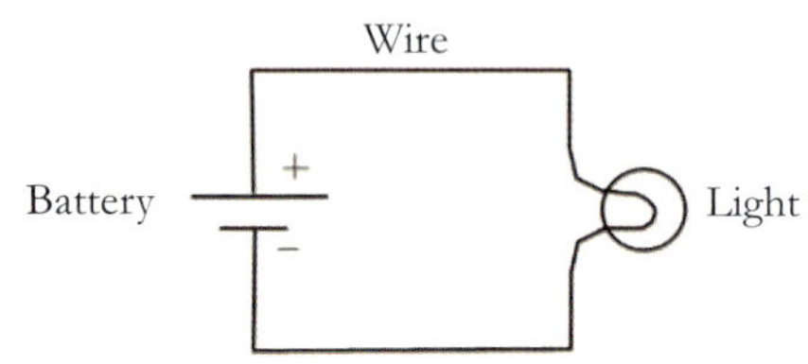

Conductors and Insulators:

Most metals are good conductors of electricity. Copper and aluminum are some of the best and easily used conductors. Not all conductors are wire shaped; sometimes a metal frame or other part of a structure is used as part of a designed circuit. But be careful. If you don't want certain metal items as part of a circuit, you'll also need insulators to block the electrical flow in areas that you don't want electrons to go. Some good insulators are made of wood, rubber, or plastic. These insulators can be placed between, around, or covering a conductor to help direct where the electrical current will flow.

Make your own conductors: Use craft sticks with holes and cover (wrap) them with aluminum foil to create electrical connectors. Then use brads, paper clips, and mini lamp sockets to complete the circuits. Use masking tape and rubber bands to hold the connections and battery in place.

SERIES Circuit

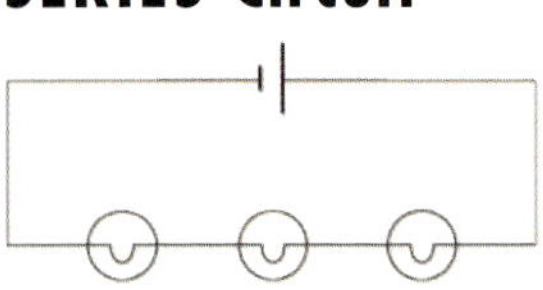

A series circuit is one continuous loop from the battery, conductors, the bulbs, and then back to the battery. To set up the series circuit, make a single path with the foil-covered craft sticks to the lamps and battery. Use brads and paper clips at the connections.

PARALLEL Circuit

A second way to set up a circuit is to have multiple paths or loops for the electrical flow from the battery. In a parallel circuit, each of the three lamps will have its own direct link to and from the battery. Make a parallel circuit with multiple loop connections with the foil-covered craft sticks, lamps, and battery. Use brads and paper clips at the connections.

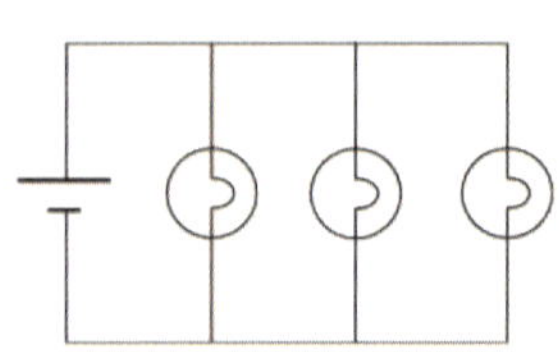

Test both circuits with a battery. Compare what happens with the lights as you remove one light at a time. Also, note the brightness of the lights in each circuit: Are lights brighter in one circuit but not the other? What happens if you light one, two, or three of the lights in a series circuit?

Activity B – Off and On

Performance Task For Youth

You will create an on/off switch for a basic circuit.

Success Indicators

Youth will be able to turn off and on a circuit they created to power a light and/or motor.

List of Materials Needed

- Robotics Notebook
- Activity Supplies
  - Copies of handout/poster of electrical circuits on page 18
  - Copies of handout/poster of electrical switches on page 20
  - Craft sticks (with holes), brass brads, wire or aluminum foil, paper clips, tape, light bulbs, and batteries
  - A base for building circuit design of pegboard, cardboard, or needlepoint board
  - Or, the circuits made from Activity A – Series/Parallel

Activity Timeline and Getting Ready

- Activity will take approximately 20 minutes.
- Divide youth into teams of two or three.
- Provide each group with the materials they can use to create an on/off switch.

Experiencing

1. Challenge youth to experiment with different designs of switches and types of material.
   a. Challenge 1: Make a single pole single throw (SPST) switch to control all the lights in the series circuit made in Activity A.

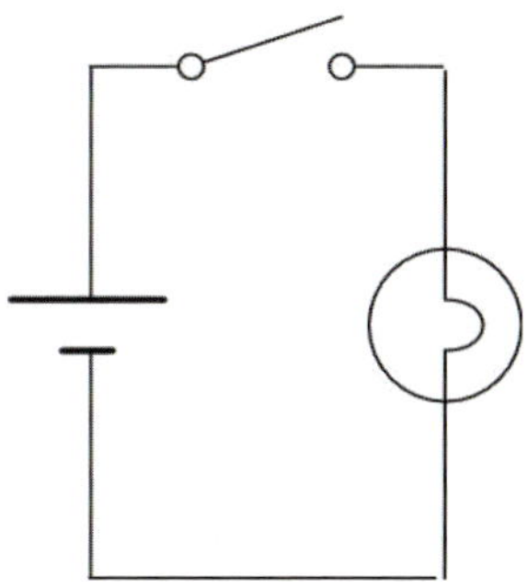

   b. Challenge 2: Make a single pole single throw (SPST) switch to control just one of the lights in the parallel circuit from Activity A.
   c. Challenge 3: Make a single pole double throw (SPDT) switch to turn on/off two different lights, one at a time.

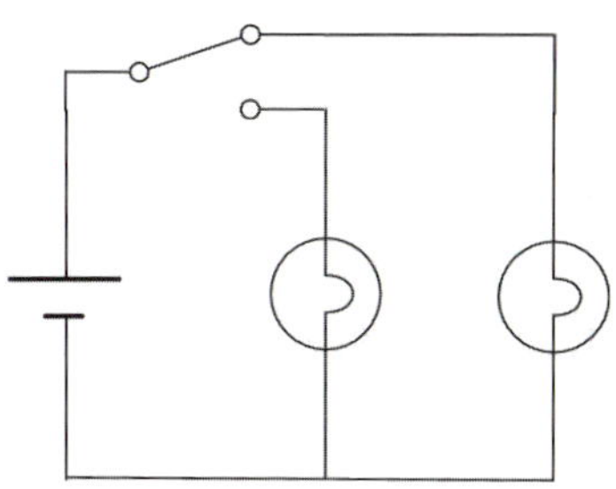

Electrical Switches

The electrical flow in a circuit may need to have a control to turn it on and off. The use of a switch or switches can provide ease of control while retaining good contacts at the connections. There are a number of different types of switches that can be made in a circuit.

Single Pole Single Throw (On/Off) Switch (SPST):

The simplest switch is a single pole single throw switch; it will open and close one circuit. Craft sticks with holes, aluminum foil, paper clips, and brass brads can be used to make an on/off switch (on – closed circuit; off – open circuit). Hint: Only cover part of the craft stick with foil.

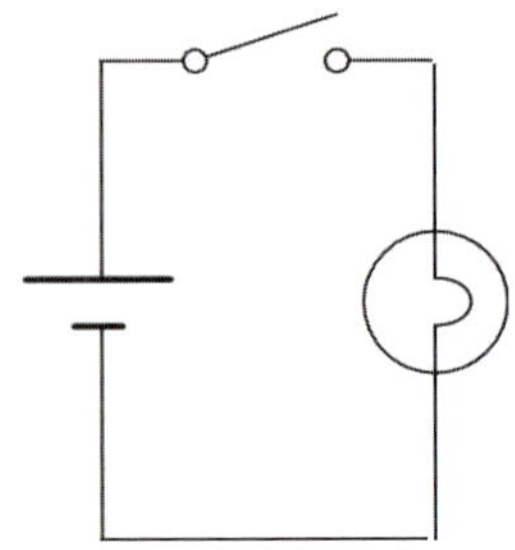

Single Pole Double Throw Switch (SPDT):

This type of switch can control two different circuits, one at a time. One circuit can be opened as a second circuit is closed.

Challenge: Try using a craft stick, one paper clip, and three brads to create a SPDT switch.

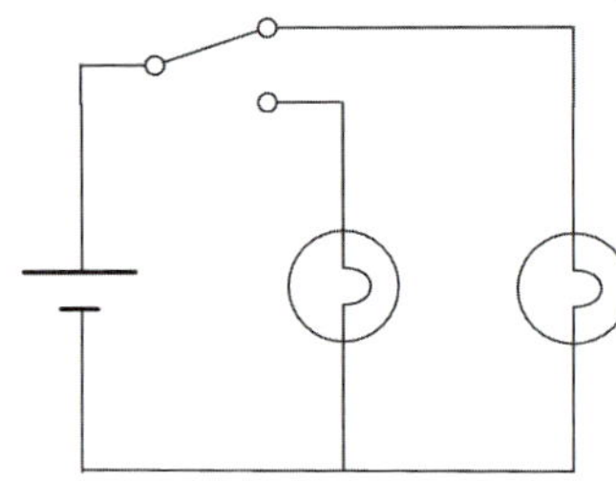

Double Pole Single Throw Switch (DPST):

This switch can operate two circuits at the same time, opening and closing them together. It is like two single pole switches next to each other.

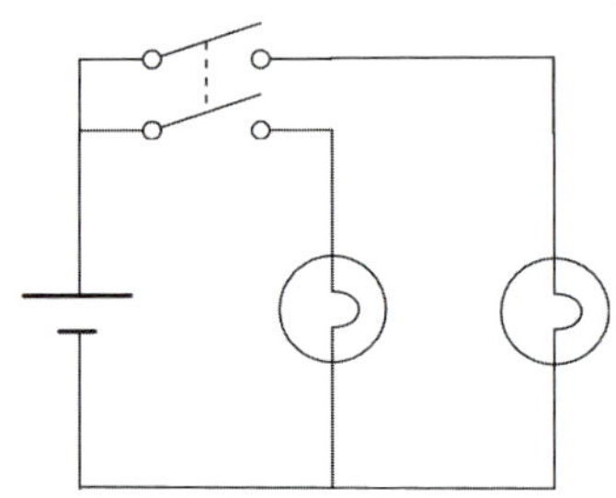

Double Pole Double Throw Switch (DPDT):

This type of switch can control a number of circuit directions, including reversing the electrical flow to change the direction of motor rotation. Below are some sample designs for DPDT switches. What design can you create for a DPDT switch?

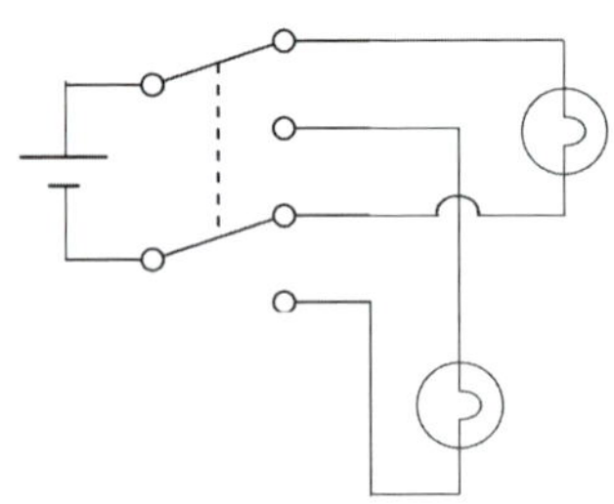

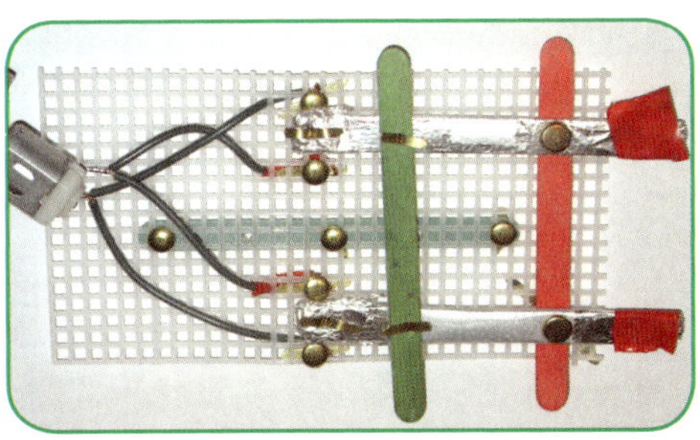

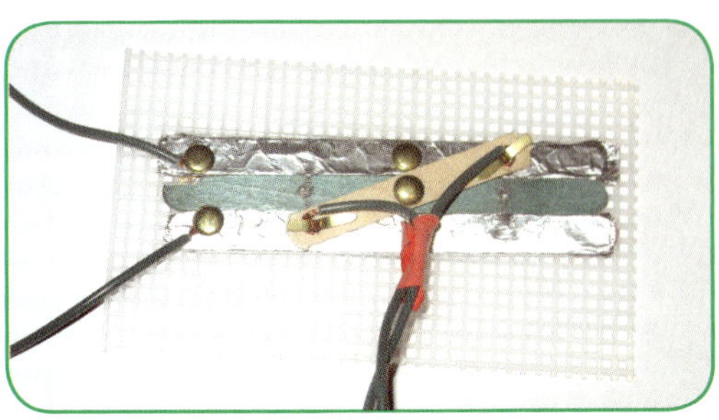

Sharing and Processing

As the faciliator, help guide youth as they question, share, and compare their observations. Have participants record their experience in their Robotics Notebook, including notes about new discoveries and knowledge. If necessary, use more targeted questions to prompt particular points.

- Which type of switch is the easiest to make? The hardest? Why?
- How could different designs of switches be beneficial to the circuit?
- Is there a way to turn the light on and off without touching the switch? (thermal, sound, motion, light)

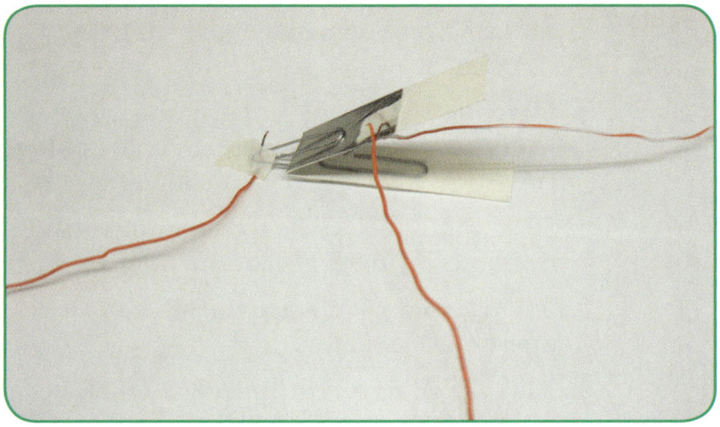

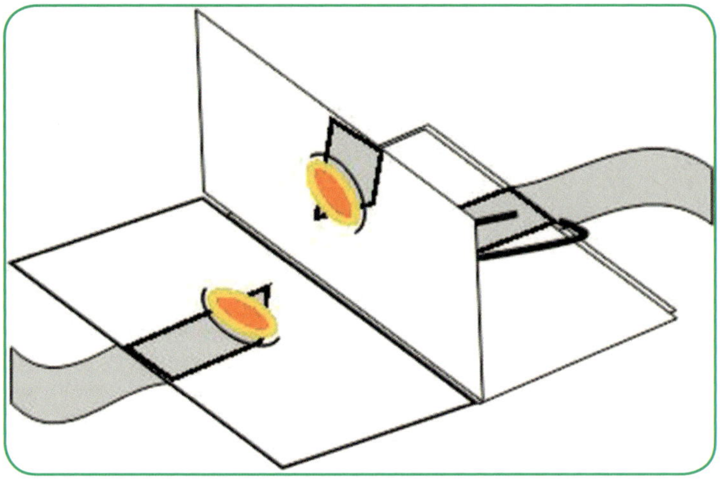

Generalizing and Applying

- Where do you use switches?
- Where could you use a switch in a robot?
- Youth can apply what they have learned in Activity C.

Activity C – Direction of Flow

Performance Task For Youth

You will experiment with circuits and control the direction of electrical flow.

Success Indicator

Youth will change the direction of electrical flow in a circuit.

List of Materials Needed

- Robotics Notebook
- Activity Supplies
  - Wire or aluminum foil and craft sticks
  - 1.5-volt batteries – AA, C, or D size
  - Toy electric motor and a wheel, gear, propeller, or cardboard disk

    Optional: holiday LED lights, if available, and parallel circuit made in Activity A
  - Copies of handout/poster of electrical circuits on page 18
  - Copies of handout/poster of electrical switches on page 20

Activity Timeline and Getting Ready

- Activity will take approximately 20 minutes.
- Divide youth into groups of three or four.

Experiencing

1. Youth will create a simple circuit with a battery, wires, and a motor.
   a. Have the members attach a wheel, propeller, gear, or a piece of tape to the motor shaft so they can observe when the motor is running.
   b. Have the participants connect the wire, battery, and motor so that the motor will run (rotate).
   c. Have youth record the direction of rotation by observing the wheel or the piece of tape.
   d. Ask the youth to change the motor's direction of rotation.
   e. Have youth record the direction of flow and how they were able to reverse the motor's direction of the rotation.
   f. Ask some sharing and processing questions about how they were able to get the rotation to change.
2. Optional: If holiday LED lights are available, use the parallel circuits made in Activities A and B, and have the youth replace one of the regular holiday mini lights in the circuit with a holiday LED light.
   a. Do all the lights light up?
   b. Change the polarity from the battery and observe the lights again. Are they on?
   c. What happens to the lights when the direction of electrical flow changes? (LED lights will only light in one direction of the electrical flow, but regular light bulbs will light either way.)

Note: LEDs are sensitive to current and are usually designed to work in circuits with a resistor to limit the current. They are being used in this situation only to help show the direction of electrical flow since they only allow flow in one direction. Use of LEDs in an electronic circuit would require calculated resistance to prevent burning out the LED.

Sharing and Processing

As the facilitator, help guide youth as they question, share, and compare their observations. Have participants record their experience in their Robotics Notebook, including notes about new discoveries and knowledge. If necessary, use more targeted questions to prompt particular points.

- What happened with the motor when the electrical flow changed direction? Why do you think this happened?
- What happened with the lights when the electrical flow changed direction? Why do you think this happened?

Compare how the direction of flow affected the light bulb, the motor, and an LED light. What did you notice with each?

Generalizing and Applying

- Where and why would direction of flow be important?
- Where would direction of flow not matter in a circuit?
- Besides electricity, where else in the world does direction of flow matter?
- Youth can apply what they have learned in Activity D.

Activity D – Forward and Reverse Design Team

Performance Task For Youth

You will plan and design a double pole double throw (DPDT) switch to change the circuit's polarity.

Success Indicator

Youth will plan, design, and share what they have learned about a DPDT switch.

List of Materials Needed

- Robotics Notebook
- Activity Supplies
  - Craft sticks, brass brads, wire or aluminum foil, paper clips, tape, light bulbs, toy motor, wheel, gear, or propeller, and batteries
  - A base for holding the parts of the circuit made of pegboard, cardboard, plastic canvas, or other materials
  - Copies of handout/poster of electrical circuits on page 18
  - Copies of handout/poster of electrical switches on page 20

Activity Timeline and Getting Ready

- Activity will take approximately 20 minutes.
- Divide youth into groups of two or three.

Experiencing

1. Display/show materials to be used for this activity. Design teams may look at – but not touch or play with – items that they will be using during the design stage.
2. Review the concepts of series and parallel circuits and types of switches with the group.
3. Give limited instructions on the design task.
4. Design and draw a circuit and a switch that can change the direction of electrical flow and the motor's direction of the rotation.

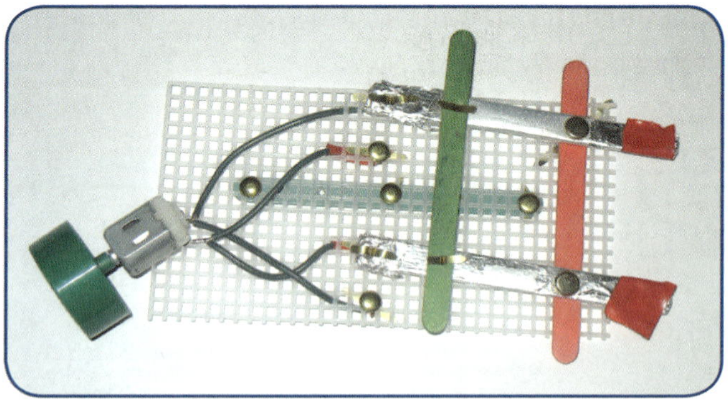

Sharing and Processing

As the facilitator, help guide youth as they question, share, and compare their observations. Have participants record their experience in their Robotics Notebook, including notes about new discoveries and knowledge. If necessary, use more targeted questions to prompt particular points.

- Have the youth share their design with everyone, allowing them to share how it works and why this particular design would be good to control a motor.
- What other material can be used to create the switch?

Generalizing and Applying

- Encourage youth to try different designs and ideas.
- Is there something that would work better or easier?
- How could you use a different type of switch to change the polarity for motor control?
- Youth will take their design and create a double pole double throw switch for a simple circuit.
- Youth will apply this knowledge by building the switch in Activity E.

Activity E – Forward and Reverse Build Team

Performance Task For Youth

You will use your Double Pole Double Throw (DPDT) design and build a working switch that you can attach to a simple circuit.

Success Indicator

Youth will create a DPDT switch that will reverse a motor's rotation.

List of Materials Needed

- Robotics Notebook
- Trunk of Junk
- Activity Supplies
  - Aluminum foil and craft sticks (with holes), 20- to 22-gauge wire (phone cord size)
  - Paper brads, paper clips, rubber bands, etc.
  - Masking, clear, or electrical tape
  - A base or backing to hold the switch components made of cardboard, plastic canvas, or another material
  - Copies of handout/poster of electrical circuits on page 18
  - Copies of handout/poster of electrical switches on page 20

Activity Timeline and Getting Ready

- Activity will take approximately 40 minutes.
- Use the same groups from Activity D, Forward and Reverse Design Team.

Experiencing

1. Allow the teams to build. Encourage them to continue testing their design.
2. Have participants record any changes they made, complications, and successes in their Robotic Notebook.
3. Have participants present their switches and allow them to demonstrate their working switches.

Sharing and Processing

As the facilitator, help guide youth as they question, share, and compare their observations. Have participants record their experience in their Robotics Notebook, including notes about new discoveries and knowledge. If necessary, use more targeted questions to prompt particular points.

- What design seemed to work the best and why?
- What are some other ways you could have built your switch?
- How might one of the switches be used in a robot?

Generalizing and Applying

- What are some real world applications of the double pole double throw switch?
- How can a double pole double throw switch work on a robotic arm? What movements can be created?
- Youth also can apply the knowledge they've gained in the next module.

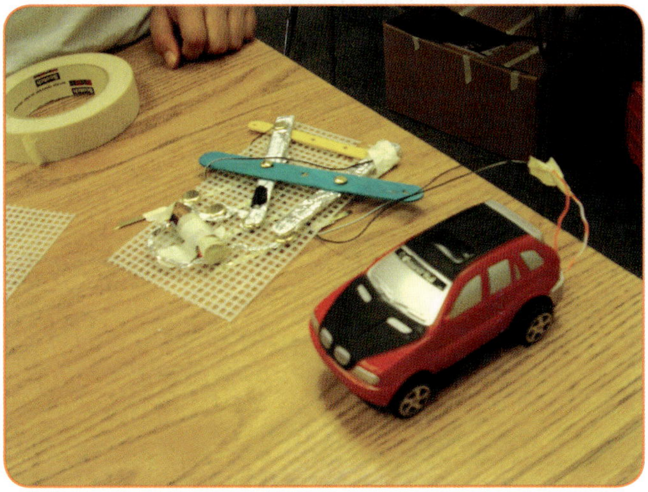

CAREER CONNECTIONS

Using Nature as an Inspiration

Engineers use science and math in order to develop safe and useful solutions to real-world problems. So where do engineers get the ideas for these creations? Often, they simply look out their window!

We live in an incredible natural world. Earth has a wonderful natural environment with highly adapted animals and plants. The arms, eyes, wings, and leaves of plants and animals have been functioning for centuries. These highly developed designs have already solved some of the problems humans are struggling with, and they serve as inspirations for engineers. Almost all engineering products imitate some aspect of nature. This idea is called "biomimicry."

For example, owls were the basis for military planes that are undetectable by radar. The locative abilities of the bat led to mobile robots with their own guidance systems. Butterflies with fluorescent wings helped redesign LED lights. The super strength of spider webs enabled sturdier materials and building designs. Red seaweed inspired resistant antibiotics.

As you can see, engineers have a whole world of natural phenomenon to study! Creatures, plants, and geography help engineers satisfy our human needs. Nature serves as a model, mentor, and measure for engineers.

- Can you think of any products in your home that may have been modeled after an aspect of nature?
- Think of a plant, animal, or location. What natural mechanisms do they have that could be useful for humans?

Adapted from: *http://www.biomimicryinstitute.org/*

Module 2: Come to Your Senses

Overview of Activities in this Module

To Learn

Activity F – Line Follower
Activity G – Keep in Touch
Activity H – Don't Buzz Me!

To Do

Activity I – Wall Follower Design Team

To Do

Activity J – Wall Follower Build Team

Note to Leader

The five senses of smell, touch, taste, sight, and hearing bring us an understanding of the world. We can use these senses to develop our skills of observation and explore the world around us. Using our senses can help us react to the changing environment – detect obstacles, escape hazardous situations, and make better choices.

How can robots make observations and change behavior without human intervention? Just as we use our five human senses, a robot can have sensors to help it react to its surroundings. Robot sensors come in a wide variety of forms and functions. They may sense light, motion, heat, sound, physical objects, moisture, and many other physical properties. Basic sensors may be mechanical, electrical, or operated by some other power source. Simple sensors can range from simple bumper switches to more complex vision cameras. More complex systems help robots sense light/dark, smooth/prickly, hot/cold, and soft/hard, etc.

To serve a function, sensors must have a purpose. This can range from reversing electrical flow, causing a motor to reverse direction, to a robotic video camera observing the road, feeding data into a computer system and directing a vehicle along a safe path.

"Line following" is a common simple navigation scheme for robots. You can control a robot's motions by placing a line on the ground and programming its sensors to follow the line. One way this scheme works is by telling the robot to turn or go straight, depending upon what it "sees."

In this module, youth will explore some basic concepts of sensors and then build a robot vehicle that can sense a barrier and reverse direction.

What you will need for Module 2: Come to Your Senses

- Robotics Notebook for each youth
- Trunk of Junk, see page 8
- Activity Supplies
  - Several long wrapping paper tubes or lengths of ¾-inch or 1-inch PVC pipe
  - Blue painter's tape (leaves less sticky residue when removing)
  - Blindfolds or eye patches (optional)
  - Sheets of newspaper
  - Braille alphabet card, cards with words written in Braille
  - Blank note cards to write down letters for words if Robotics Notebook is not available
  - Pieces of wood, plastic, drilled craft sticks
  - Buy a cheap motorized toy car or four wheels per group and other materials needed to build a car, including gears, motors, axles, and batteries (AA, C, or D)
  - Various items such as straws, string, rubber bands, nails, tape, glue, eye hooks (optional)
  - Purchase or make a steady hand/buzz wire game. See the 4-H Robotics website at *www.4-H.org/curriculum/robotics* for more details and suggestions.

Timeline for Module 2: Come to Your Senses

Activity F – Line Follower

- Activity will take approximately 20 minutes.
- Divide youth into groups of two or three.

Activity G – Keep in Touch

- Activity will take approximately 20 minutes.
- Divide youth into groups of three.

Activity H – Don't Buzz Me!

- Activity will take approximately 20 minutes.
- Divide youth into two or three groups based on the number of buzz wire games available.

Activity I – Wall Follower Design Team

- Activity will take approximately 20 minutes.
- Divide youth into groups of three to four.

Note: If a car has not been built, youth will need to design and build a car along with the switch. (Refer back to Level 2, Module 3, Activities K and L). The teams could use a rover like the "Es-Car-Go" they built in Level 2 or a toy electric car from a "dollar store" that could be modified to add the switch.

Activity J – Wall Follower Build Team

- Activity will take approximately 30 to 50 minutes, depending on if the vehicles are ready to use or have to be built.
- Use the same groups from Activity I.

Focus for Module 2: Come to Your Senses

Big Ideas

- Simple sensors help control robot behavior.
- Many different types of sensors are available to provide information to the robot.
- Sensors enable robots to work autonomously (without human control).

NSE Standards

- Systems, order, and organization
- Abilities of technological design
- Understanding about science and technology

Performance Tasks For Youth

You will follow a given line and learn how a visual sensor works on a robot.

You will use the sense of touch to read Braille code.

You will navigate two different size loops through a wire course to determine sensitivity and accuracy.

You will plan and design a robot around the given wire diagram that will follow a wall.

You will build a robot that will travel around an object or wall using a sensor for control.

STL

- Relationships and connections
- Apply the design process
- Social effects of technology
- Information and communication technologies

SET Abilities

- Communicate/Demonstrate
- Draw/Design
- Build/Construct
- Observation

Life Skills

- Keeping records
- Cooperation
- Teamwork

Success Indicators

Youth will understand how a robot uses a visual sensor.

Youth will be able to decipher Braille code to read and record words.

Youth will be able to describe how sensitivity and accuracy are important in designing robots.

Youth will share a design they have for a robot that is able to follow a wall or similar objects.

Youth will use their designs to build a robot that can sense and follow along walls or objects.

Activity F – Line Follower

Performance Task For Youth

You will follow a given line and learn how a visual sensor works on a robot.

Success Indicators

Youth will understand how a robot uses a visual sensor.

List of Materials Needed

- Robotics Notebook
- Activity Supplies
  - Several long wrapping paper tubes or lengths of ¾-inch or 1-inch PVC pipe
  - Blue painter's tape (leaves less sticky residue when removing)
  - Newspaper
  - Pencil or felt pen

Activity Timeline and Getting Ready

- Activity will take approximately 20 minutes.
- Obtain all necessary materials.
- Divide youth into small groups of two or three.
- Before the start of the session, make a winding line about 2 feet long on the newspaper using the blue tape.
- If possible, make enough newspapers with lines so that each group has its own paper and line. Try to have each line form a different pattern or at least more than one pattern.
- Create binoculars by putting together pairs of tubes that are around 3 feet long. Shorter tubes can be taped end to end to make longer tubes to fit the height of the youth. The binoculars should reach above waist level so youth can bend over to look down the tubes at the line on the floor.
- The top part, where the youth will look down, should be set at a distance to closely match eye spacing. Also create a "blinder" on the outside of the tubes so the youth are not allowed to sneak a peek at the line.
- Make some of the binoculars with different tube spacing on the end closest to the floor. Tape or fasten a felt pen or pencil, pen, or marker to the end of all the tubes so they hold the bottom of the tubes about an inch off the floor and draw lines as the tubes are moved to follow the line. (Caution: If using felt pens that "bleed," protect the floor with extra paper, cardboard, or other backing. Also, you may have to tape the newspaper down so that it will not move as the markers follow the line. Another option would be to use a driveway, playground, or sidewalk, and have chalk on the ends of the tubes.)

Experiencing

1. Have the youth gather around the lines and explain their roles. One person at a time will be the sensor using the "binoculars." The sensor will look down the tubes and try to follow the line. The sensor must keep the line between the two bottom openings of the tubes, so the sensor should start at the beginning of the line and look down through the tubes. The sensor will move his or her head and the tubes left and right while trying to keep the blue tape line between the two tubes as he or she walks forward (baby steps like crawling slowly).
2. After the youth experience how sensors work and how to keep the line between the two tubes, have them exchange their binoculars with another youth to see how different tube spacing will affect their ability to follow the line. The markers will trace the movement of the sensor and will allow the youth to see how accurate they were with the line following. (Using markers of different colors with the various binoculars will help to track the patterns of sensitivity in following the line.)
3. Youth should try all the different distances of the tubes to see which was more accurate and if the spacing affected the time needed to follow the line.

Sharing and Processing

As the facilitator, help guide youth as they question, share, and compare their observations. Have participants record their experience in their Robotics Notebook, including notes about new discoveries and knowledge. If necessary, use more targeted questions to prompt particular points.

Have youth discuss what they think a visual sensor on a robot does and how it helps the robot function.

- Are visual sensors the only sensor a robot can have? Have youth discuss the different types of sensors and their uses on a robot.
- How are accuracy and precision related? Is one more important than the other or should a robot be able to have both skills?

Generalizing and Applying

Why was changing the distance between the tubes important? How did it affect the ability to follow the line?

- Have youth discuss other types of visual sensors, and how they operate.
- Sensors are important to a robot's function. What sensors do humans have and how do we use them in our daily lives?
- Youth can apply what they have learned in Activity G.

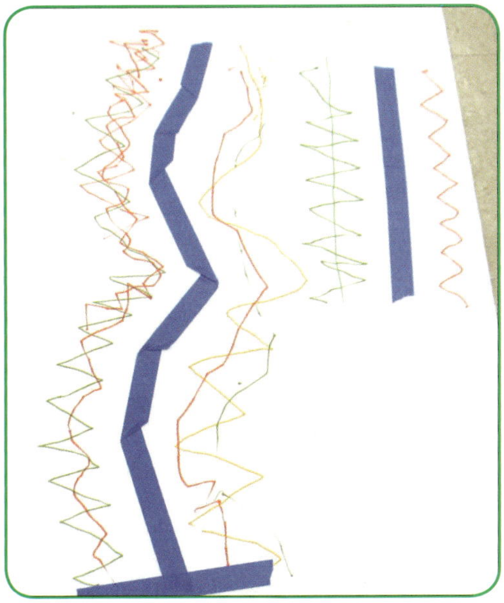

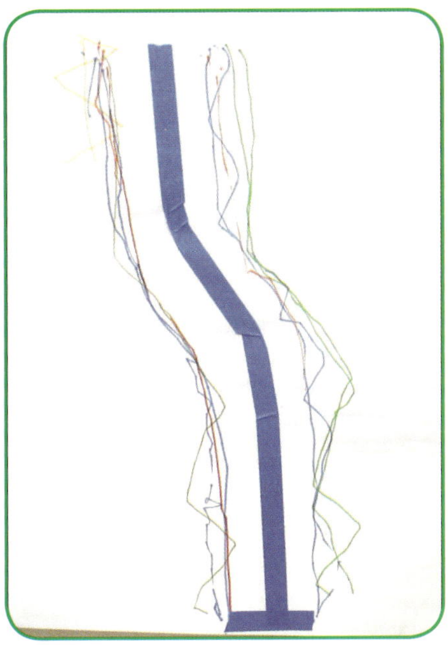

Activity G – Keep in Touch

Performance Task For Youth

You will use the sense of touch to read Braille code.

Success Indicators

Youth will be able to decipher Braille code to read and record words.

List of Materials Needed

- Robotics Notebook
- Activity Supplies
  - Blindfolds (one for each team)
  - Braille alphabet cards, cards with words written in Braille
  - Blank note cards to write down letters for words if Robotics Notebook is not available

Activity Timeline and Getting Ready

- Activity will take approximately 20 minutes.
- Divide youth into groups of three.
- Provide each team with supplies.
- Use the Robotics Notebook or make copies of pages 36-37 for each group.
- Have word cards made in Braille or make your own using glue drops, sticky "rolling eye" dots, or sticky label dots on the code grid as shown. Make a copy of the pages for each team. Apply the dots (bumps) in the code locations to spell out letters and words. The first one is already coded; you just need to apply dots so they can be felt. Make at least two or three more words in Braille. Some school districts may have a Braille printer that you could access for a printout of words in Braille.

Note:

Braille is not a language but a code to read other languages. It is a code that is read by touching a group of bumps or dots. This code was developed in the 1800s primarily for reading and writing for individuals who are blind or visually impaired. Braille has been improved and expanded over the years, including use in many languages, math, music, and more. The basic design of a Braille cell is a six (6) dot system. The cell is two columns by three rows with many combinations of dot patterns representing letters and numbers. The dot spaces are numbered for reference: 1, 2, 3 are in the first column top to bottom, and 4, 5, 6 are in the second column top to bottom.Technology has helped improve production of books and other materials in Braille. Special printers attached to computers can emboss (print) Braille, books in Braille can be read and translated into languages, and much more. In fact, there are Braille computer fonts that are not embossed. This is "Junk Drawer Robotics" in Braille: 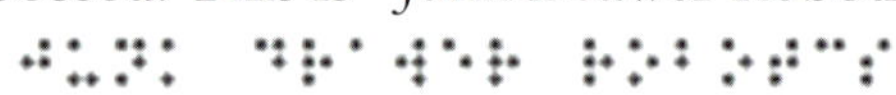

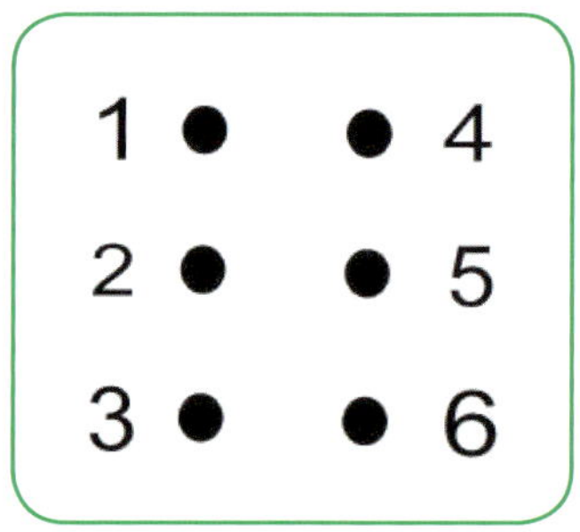

In this activity, youth will explore touching (reading) the dot code, de-coding it to letters, and then outputting it into words. Discussion about how a sensor, computer, or robot could use or gather this information will help youth understand sensor use and applications in robotics.

Experiencing

1. After a brief discussion about what participants know/don't know about Braille, have each member of the team take a role: *sensor*, *processor*, and *output*. If the team has only two members, combine *processor* and *output*.
    a. Directions for *sensor (input)* - The *sensor* should put on a blindfold and be given cards with words in Braille. He or she should touch the bumps and describe the pattern to the *processor*, one letter at a time.
    b. Directions for *processor* - The *processor* should use the grid chart to mark down the location of dots or bumps. Then, he or she should use the Braille alphabet code chart to figure out the letter being described by the *input* and relay that letter to the *output*.
    c. Directions for *output* - The *output* should keep track of the letters received from the *processor* and write them on note cards to reveal the final message.
2. If time allows, rotate team roles and repeat with additional Braille words.

Sharing and Processing

As the facilitator, help guide youth as they question, share, and compare their observations. Have participants record their experience in their Robotics Notebook, including notes about new discoveries and knowledge. If necessary, use more targeted questions to prompt particular points.

- How is reading Braille words different from regular printed words?
- Did you find problems in reading Braille?
- What are some advantages of using Braille?
- Could you use other senses to "read" words? Sight — lights in place of the dots; hearing — use Morse Code for messages sent using sound or light.

Generalizing and Applying

- How could a robot use Braille?
- How could a robot use Morse Code?
- What new roles or modifications to the three roles could be added to make this exercise easier?
- Youth can apply what they have learned in Activity H.

Sensor

Below, create words for *"sensor"* to touch and describe to *"processor."*

1. 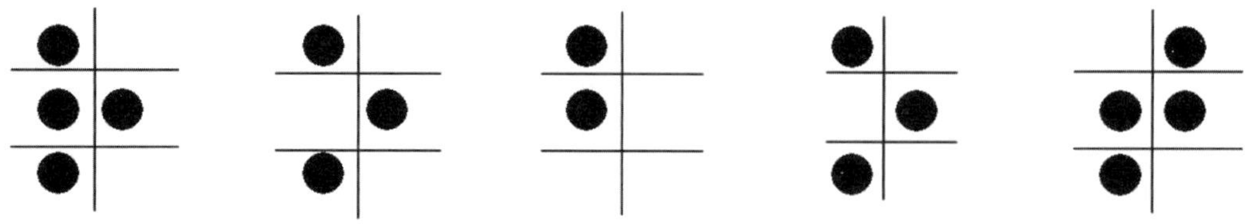

2.

3.

Output

"Output" write down letters when told by *"processor."*

1. ______ ______ ______ ______ ______

2. ______ ______ ______ ______ ______

3. ______ ______ ______ ______ ______

Processor Code deciphering area for Braille letters and words

Draw dots where *"sensor"* indicates for each letter. Do one word, then use the code chart below to de-code letters one at a time and tell "output" the letter. Repeat for the next word.

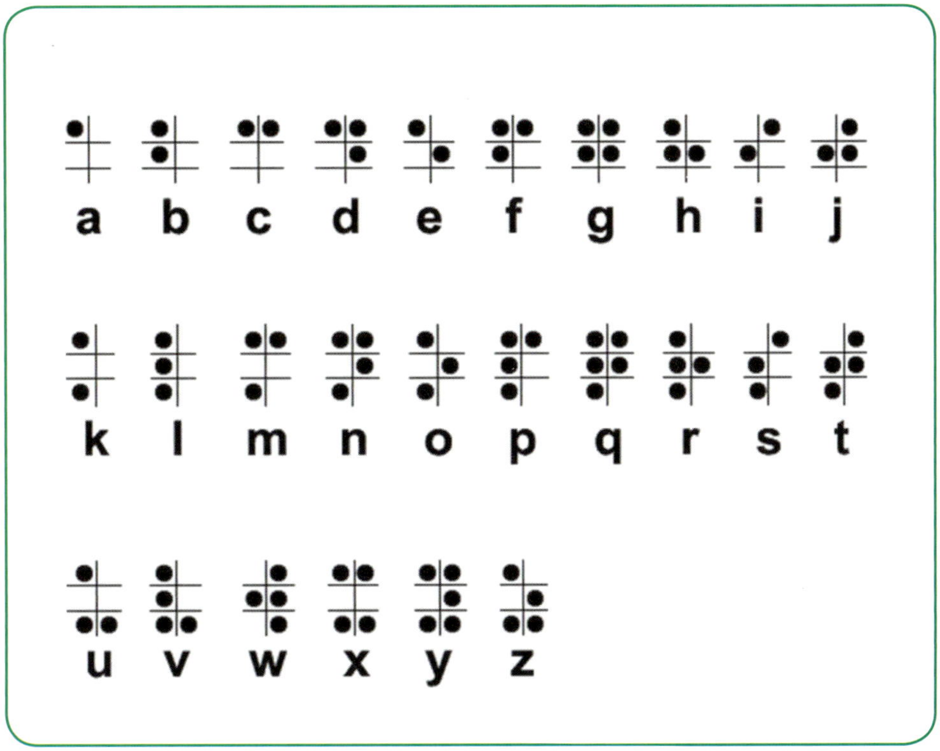

1.

2.

3.

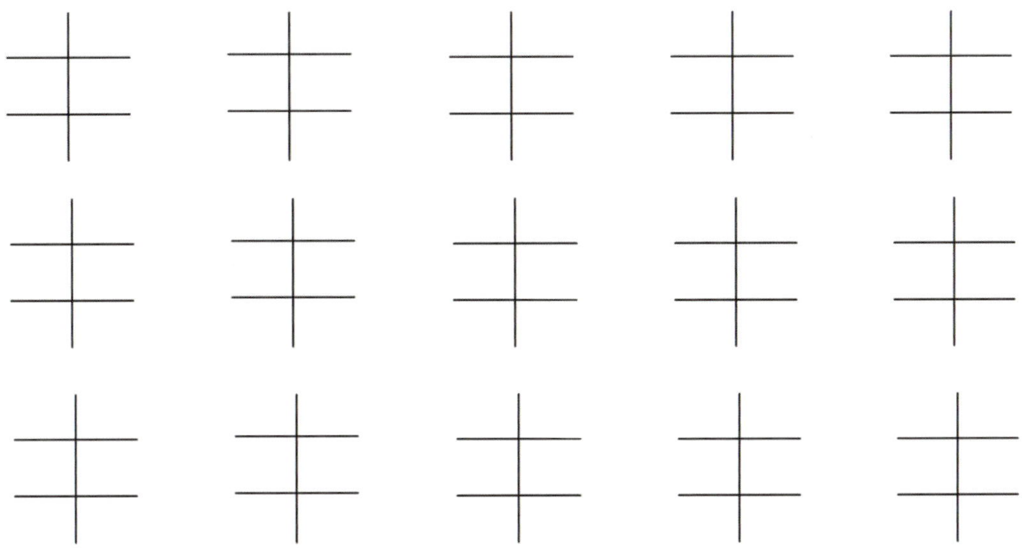

Activity H – Don't Buzz Me!

Performance Task For Youth

You will navigate two different size loops through a wire course to determine sensitivity and accuracy.

Success Indicator

Youth will be able to describe how sensitivity and accuracy are important in designing robots.

List of Materials Needed

- Robotics Notebook
- Activity Supplies
  - Purchase or make a steady hand/buzz wire game for each group. See the 4-H Robotics website at *www.4-H.org/curriculum/robotics* for more details and suggestions about making your own game.
- Optional: Copy of the Don't Buzz Me chart on page 40 for each group.
- Materials needed to build your own:
  - Stiff wire (12- or 14-gauge, coat hanger, or similar)
  - A battery (AA, C, or D) and flashlight or holiday light bulbs, small electric buzzer, or toy motor
  - Electrical tape
  - Plywood, plastic, or cardboard for a base, about 8 inches by 10 inches in size
  - 18- to 20-gauge electrical wire

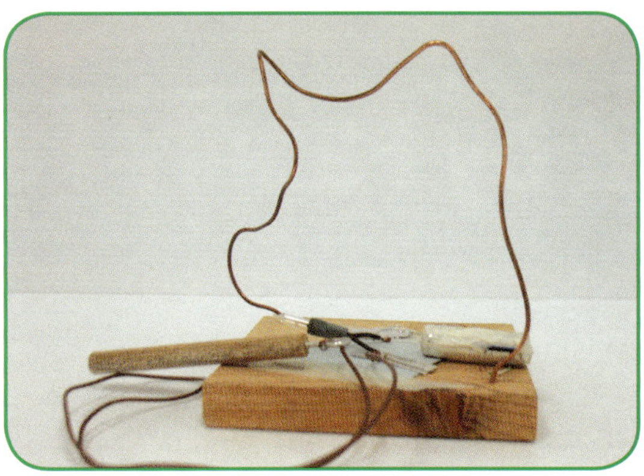

Activity Timeline and Getting Ready

- Activity will take approximately 20 minutes.
- Divide youth into small groups of three to five.
- Purchase or create "Don't Buzz Me!" activity boards for teams to share in the activity. Review the instructions for creating the boards on the 4-H Robotics website at *www.4-H.org/curriculum/robotics.*
- The activity board will consist of a stiff wire bent into a free-form shape. The bent wire will be attached to a battery and a wand with a loop, like an eye screw or "J" hook. The loop will be attached to a lamp, buzzer, or another item and back to the battery. If the loop touches the bent wire, it will close the circuit and light the lamp or sound the buzzer. This activity is designed to use at least two different size loops. The loops represent the sensitivity of a sensor.
- Note: If you purchased a buzz wire game, you may have to create additional loops of different sizes for this activity.

Note

In this activity we will be looking at the concepts of sensitivity, accuracy, and precision. These terms have definitions and uses. Some are based on human interactions, others on math or statistics. For our robots, let's start by looking at accuracy, which is how close something is to the real or intended location or value.

In a robot arm, the accuracy in moving to a specific X, Y, Z coordinate would depend on how close it can come to that target. Precision is related to repeatability of each move to that same target. In this case, can the robot arm move back to the same X, Y, Z coordinate 10 times, 100 times, or more? While the accuracy and precision of robot movement may be, in part, related to the power system (electrical, pneumatic, hydraulic, etc.), other times it will rely on the sensitivity of sensors and actuators. In this case, sensitivity is the ability to register or detect small differences in physical changes or inputs.

A robot light sensor that can detect white and black colors is good, but a sensor that can detect white, light gray, gray, dark gray, and black is more sensitive and might be better. A more sensitive sensor can provide more data input to the robot and might allow it to be more accurate and precise. On a side note, many of the things that we and robots sense are analog (a graduated form with lots of variations like a grayscale, temperature ranges, sound levels, etc.), but a computer normally uses digital computation, so it may need to be translated from analog to digital and maybe back from digital to analog on output. We will address this in Module 3, It's Logical.

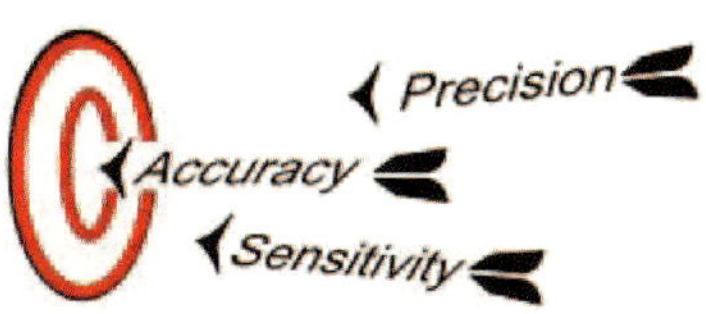

Experiencing

Step 1

Have youth try to navigate the wire course using the larger loop with the stiff wire going through the center of the loop. They should attempt to not make the motor or buzzer sound as they move the loop from one end of the wire to the other end. Have team members record in their Robotics Notebook the number of times the motor/buzzer was activated. Also record the time it took to complete the navigation. Repeat the loop activity three times for an average per person.

Step 2

Repeat Step 1 using the smaller loop. Have youth chart the results for both the large loop and the smaller loop in their Robotics Notebook.

Step 3

In addition, record both the number of buzzes and the time for each try on a poster for each person doing the buzzer activity.

Step 4

Use the chart to compare and discuss the following about each person, the loop size, or other criteria: 1) Accuracy, 2) Precision, and 3) Sensitivity.

Sharing and Processing

As the facilitator, help guide youth as they question, share, and compare their observations. Have participants record their experience in their Robotics Notebook, including notes about new discoveries and knowledge. If necessary, use more targeted questions to prompt particular points.

- Why does the buzzer or light turn on when the loop touches the wire?
- What happened when you used a larger loop? A smaller loop?
- How does this relate to the sensitivity of a sensor?
- What made this task easy or hard to do?
- What would make it easier?

Generalizing and Applying

- What advice would you give others trying to complete this activity?
- How would you design a robot to move the loop around the wire?
- What are some other scales of sensitivity that you have seen on other sensors? (Thermometer? Noise rating? Street lights?)
- Youth can apply what they have learned in Activity I.

Don't Buzz Me!

Use the chart to record the results of sensitivity using two different size loops or hooks on the wire game.
Repeat three times for an average.
Record the number of Buzzes and the overall time to pass the loop from one end of the wire to the other end.

Person and loop size	Number of buzzes		Time to complete
	Buzz marks	Total	Time
Name / large loop			
1			
2			
3			
average			
Name / small loop			
1			
2			
3			
average			

Activity I – Wall Follower Design Team

Performance Task For Youth

You will plan and design a robot around the given wire diagram that will follow a wall.

Success Indicator

Youth will be able to share a design they have for a robot that is able to follow a wall or similar object.

List of Materials Needed

- Robotics Notebook
- Trunk of Junk
- Activity Supplies
  - Index cards, paper clips, and masking tape
  - Aluminum tape, aluminum foil, wires, and paper or plastic cups
  - 1.5- to 3-volt toy motors (2 for each robot)
  - Batteries (AA size)
  - Wheels or other items to put on motor shafts
- Optional: Copies of handout/poster of instructions on page 45

Activity Timeline and Getting Ready

- Activity will take approximately 20 minutes.
- Divide youth into groups of three or four.
- Handout of wire diagram on page 45.

Note: If a car has not been built, youth will need to design a car along with the switch. (Refer back to Level 2, Module 3, Activities K and L). The teams could design a rover like the "Es-Car-Go" they built in *Robots on the Move* or use an inexpensive toy electric car that could be modified to add the switch.

Experiencing

1. Have youth look over and review the wire diagram handout. Have them discuss how the circuit works and what they think the circuit can be used for. The main components of the circuit are a SPDT (Single Pole Double Throw) switch that will function as the sensor, two toy motors, and a battery. The youth have already built a SPDT switch in Activity B, and should already be familiar with its functions. The SPDT switch can control which motor is turned on and which motor is off by creating a lever or arm to operate the switch. The switch should be made so that it is normally in one position, but springy so the arm can move it to the second position and when let go, it will return to the first position.
2. After youth have discussed the wire diagram, and understand its function, allow the youth to begin designing their robot. The robot should be autonomous — the robot should be able to follow a wall without a person in control of the movement of the robot. The robot should be able to follow along a wall and move forward while maintaining a close distance to the wall.
3. Have each Design Team draw their design in their Robotics Notebooks.

Hint: Let youth discover their own ideas and give them plenty of time. If they encounter difficulty, you can share the following:

- If the motors have wheels attached, correct rotation, and position on the "cup" or frame, the robot should move forward in a turning direction either to the left or right.
- When the robot turns into the wall, the switch is moved or activated by touching the wall. It should turn off the first motor and turn on the second motor to move the robot forward but now turning in the opposite direction right or left and away from the wall.
- Then, as it turns away from the wall, the switch will lose contact and return back to the first setting, causing the robot to turn back toward the wall and repeating the sequence in a side-to-side forward motion.

Sharing and Processing

As the facilitator, help guide youth as they question, share, and compare their observations. Have participants record their experience in their Robotics Notebook, including notes about new discoveries and knowledge. If necessary, use more targeted questions to prompt particular points.

- Have the groups share their design.
- What are some ways a robot can sense when it is time to change direction?
- What type of sensors are used in elevator doors? In a garage door? With traffic lights?

Generalizing and Applying

- Encourage youth to try different designs and ideas.
- Is there something that would work better or easier?
- How could you use a different type of switch to change the polarity for motor control?
- Youth will take their design and create a wall following robot.
- Youth will apply this knowledge by building the switch in Activity J.

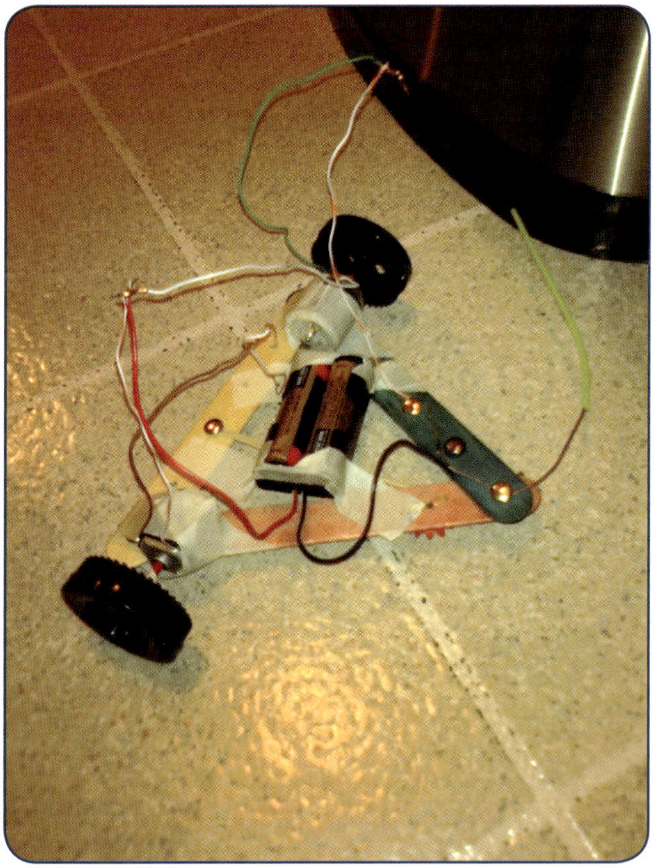

Activity J – Wall Follower Build Team

Performance Task For Youth

You will build a robot that will travel around an object or wall using a sensor for control.

Success Indicator

Youth will be able to use their designs from the previous activity to build a robot that can sense and follow along walls or objects.

List of Materials Needed

- Robotics Notebook
- Trunk of Junk
- Activity Supplies
  - Index cards, paper clips, and masking tape
  - Aluminum tape, aluminum foil, wires, and paper or plastic cups
  - 1.5- to 3-volt toy motors (2 for each robot)
  - Batteries (AA size)
  - Wheels or other items to put on motor shafts
- Handout of wire diagram
- Toolbox
  - Hand drill and bits
  - Hacksaw
  - Pliers, scissors, punches, if needed
- Optional: Copies of handout/poster of instructions on page 45

Activity Timeline and Getting Ready

- Activity will take approximately 30 to 50 minutes.
- Use the same groups from Activity I.

Note: If a car has not been built, youth will need to build a car and switch (Refer back to Level 2, Module 3, Activities K and L). The teams could design a rover like the "Es-Car-Go" they built in Robots on the Move or use an inexpensive toy electric car that could be modified to add the switch.

Experiencing

1. Youth will build a robot using the wire diagram provided and designed in Activity I.
2. The robot should be able to move forward while it moves left and right as it follows around a wall or object. The robot should maintain a close distance to the wall and be autonomous. (Robot should be able to follow a wall without a person controlling the movement of the robot.)
3. Have each group test their robot and redesign it, if necessary.
4. In their Robotics Notebook, youth should make notes on any changes, modifications, or suggestions they've made to their design, as well as any other notable discoveries.
5. Optional: Youth can have a friendly challenge to see whose robot can make it around the room the fastest while still following closest to the wall.

Sharing and Processing

As the facilitator, help guide youth as they question, share, and compare their observations. Have participants record their experience in their Robotics Notebook, including notes about new discoveries and knowledge. If necessary, use more targeted questions to prompt particular points.

- Have the groups tell how they built their robot.
- Ask the groups to explain how their robot takes the sensor readings.
- Are there other ways a robot could sense?
- Have youth discuss robots they've seen that have the same properties as their robot. Are there robots like this in real life? Where can these robots be used and for what functions?
- What happens if the robot is going too fast or too slow?

Generalizing and Applying

- Where would this technology be useful?
- Can you make the robot move in different direction, like turning or following a line? How?
- Youth can apply the knowledge they have gained in Module 3.

Create a Wall Following Robot

Use or modify this SPDT motor control circuit to create, design, and build a robot that will follow along a wall or other object. Use simple items like a paper cup, battery, and two toy motors with wheels. Make a switch from note cards, paper clips, aluminum foil, or tape. Try for a simple limited parts robot that will follow the wall using its own sensing and control without help from a human.

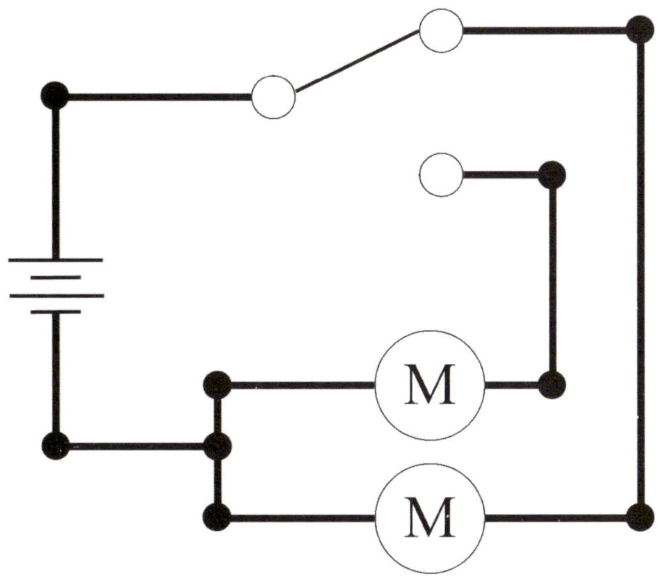

Module 3: It's Logical

Overview of Activities in this Module

To Learn

Activity K – It's About Time
Activity L – Logic: AND, OR, NOT!
Activity M – Analog Versus Digital
Activity N – Components

To Make

Activity O – Breadboard Build Team

Note to Leader

Throughout human history, man has devised and used many counting systems. The now-common base 10 systems may have stemmed from the fact that we have 10 fingers or digits. Other systems, like base 12, are still evident in many parts of our lives, such as 12 inches in a foot, a 12-month calendar, and the two 12-hour cycles in our day.

While a base 10 system, like the metric system, can be easy for humans to use, computers and other electronics struggle to process so many digits. What would be an easier counting system for an electronic computer? A system that relies on two simple functions — the "off and on" status of electric flow — would be much more practical. Putting this two-digit design into practice allows circuits in a computer to perform counting, addition, multiplication, logical decisions, and other mathematical equations to represent words and symbols, record input, provide output, and help solve problems.

This system of counting and logic is known as the binary or base 2 system, and it uses only two digits (1 and 0). These digits can represent on or off, true or false, yes or no, high or low, all as a function of the status of an electric current. Since the writing of a number in binary can be confusing to humans (18 in base 10 would be written as 10010 in base 2), programmers and software developers sometimes use a grouping of binary numbers or bits to create a binary word size (4 bits = Nibble and 8 bits = Byte). When programming and designing, programmers may use a translated base 8 or base 16 (Hexadecimal) in written code, although the computer is still using base 2.

Why is understanding binary systems important when working with robotics? In the study of robotics, specifically in mechatronics, robot control is essential. In *Junk Drawer Robotics* Levels 1 and 2, youth explored the mechanical side of robotics: levers and arms, gears and motors, pneumatics and propulsion, and electricity and magnetism. In the beginning modules of this level, youth explored senses and basic switches for mechanical control of a robot. While these mechanics are needed to maintain basic control of a robot, more advanced functions — including direct control, acceptance of input data, and specific outputs — are needed to create a more beneficial and sophisticated robot.

In this module of *Junk Drawer Robotics*, youth will be studying how the binary system is used in computer control. Youth will explore how timers are used in recording data and how logic gates help create specific results. They will be able to recognize basic electronic components, and devise a real control circuit with components on a breadboard, a board on which mounted components are used to make an experimental arrangement.

Much of daily life consists of interactions that are on a variable continuum, yet the binary system works in discrete steps. Therefore, these activities also will highlight the relationship between analog and digital inputs and outputs.

What You Will Need for Module 3: It's Logical

- Robotics Notebook for each youth
- Trunk of Junk, see page 8
- Activity Supplies
  - Copies of counting and timing charts on page 53
  - Copies of logic gate operations cards on pages 57-61
  - Samples of analog and digital interaction situations on pages 64-66
  - Copies of role cards for circuit components on pages 69-71
  - Chairs for every youth
  - A few small tables
  - Index cards and tape or sticky notes
  - Felt markers
  - Radio, recorder, iPod®, CD player, or other musical listening device
  - String, yarn, or cord, in various lengths
  - Ping-pong or other small, nonbreakable balls (20-30 or more)
  - Large container with limited size opening that can hold 20 to 30 ping-pong or similar balls
  - Two smaller containers that will only hold about five ping-pong balls
  - Two or more plastic spoons
  - Note cards (approximately 5 inches by 8 inches)
  - Clock or watch for recording time
  - An open area to move around in while playing a game
  - A small electronics experiment kit or individual components (resistors, LEDs, light bulbs, capacitors, IC chips)
  - Small, prototyping breadboard (solder-less breadboard)
  - Battery (9 volt)

Timeline for Module 3: It's Logical

Activity K – It's about Time

- Activity will take approximately 20 minutes.
- Divide youth into groups of six.
- For every group, place four chairs in a line and place a card with the following numbers on each chair:
- The chair farthest to the right is "1."
  Second, "2"
  Third, "4"
  The chair farthest to the left is "8."

Activity L – Logic: AND, OR, NOT!

- Activity will take approximately 20 minutes.
- Divide youth into groups of three or four.
- Set up the chairs and string according to the diagram.

Activity M – Analog Versus Digital

- Activity will take approximately 20 minutes.
- Divide youth into groups of 10 or more.
- Print cards from Activity L: inputs, outputs, and logical operators.

Activity N – Components

- Activity will take approximately 30 minutes.
- Use groups of at least four, up to the total group of youth. If using more than one group, each group will need its own containers and other supplies.

Activity O – Breadboard Build Team

- Activity will take approximately 40 minutes.
- Use teams of two to four to construct circuits.

Focus for Module Module 3: It's Logical

Big Ideas

- While humans typically use a base 10 number system, digital systems are more efficient with a base 2 system (on/off).
- Logical operators, used in Boolean logic, form the basis of all modern digital electronics.
- Basic electronic circuit design, including components of batteries, wire conductors, resistors, lights, capacitors, diodes, transistors, and switches.

NSE Standards

- Systems, order, and organization
- Abilities of technological design
- Understanding about science and technology

Performance Tasks For Youth

You will learn about base 2 binary counting and timing as related to electronic circuits.

You will learn about logical operators, including AND, OR, and NOT.

You will apply logical circuits in three real-world settings.

You will learn about electrical components including batteries, wire conductors, resistors, and capacitors.

You will rebuild electronic circuits you've built in previous activities.

STL

- Core concepts of technology
- Development and use of technology
- Information and communication technologies

SET Abilities

- Design Solutions
- Build/Construct
- Invent/Implement Solutions

Life Skills

- Keeping records
- Problem solving
- Sharing

Success Indicators

Youth will be able to read and write a number in binary and describe why timers are used in electronic circuits.

Youth will be able to demonstrate understanding of common logical operators.

Youth will be able to compare and describe both analog and digital.

Youth will role play an electrical circuit and act out batteries, wires, resistors, and capacitors.

Youth will create working electronic circuits using a solder-less breadboard.

Activity K – It's About Time

Performance Task For Youth

You will learn about base 2 binary counting and timing as related to electronic circuits.

Success Indicators

Youth will be able to read and write a number in binary and describe why timers are used in electronic circuits.

List of Materials Needed

- Robotics Notebook
- Activity Supplies
  - Four chairs for every group
  - Index cards and tape or sticky notes
  - Felt pens
  - Radio, recorder, iPod®, CD player, or other musical listening device
  - Copies of counting and timing charts on page 53 for each group

Activity Timeline and Getting Ready

- Activity will take approximately 20 minutes.
- Divide youth into groups of six.
- For every group, place four chairs in a line and place a card with the following numbers on each chair:
- The chair farthest to the right is "1."

  Second, "2"

  Third, "4"

  The chair farthest to the left is "8."

Experiencing

1. Pose these questions to the full group:
   a. Ask youth to describe some different counting systems that we use today. (metric system, calendar months, time, degrees in a circle, seconds in a minute, inches in a foot, etc.)
   b. If not mentioned by the group, ask what type of counting system a computer might use.
   c. How could a system, such as a computer, use electricity to represent digits? (If this question is not answered, leave it for later, as the activities will help provide the answer.)
   d. We use our hands and fingers to count to 10. How do we count to 100 or to 587?
   e. How could you count to 10 if you only had two characters (digits) to use in counting?

2. Counting in binary — Divide the group into teams of six. Have four members of the team sit in one of the four chairs that have been marked as 1, 2, 4, and 8. Explain that they now represent the number on their chair (1, 2, 4, or 8) and they are a one binary digit called a bit (short for Binary Digit). As a group, they will create a four-bit binary word called a Nibble. As a Nibble, they will be able to represent the numbers from 0 to 15. The remaining team members will serve as a recorder and a reader.

a. Each member represents zero (0) when sitting and one (1) when standing. Each chair represents a different placing (1, 2, 4, and 8). The group's answer will be the summation of the place numbers represented by standing youth.

b. Using the counting chart, the team *recorder* will record positions for each of the numbers, beginning with zero. With all the members/bits sitting, the answer will equal zero (0+0+0+0=0), and with all the bits standing, the answer will equal 15 (8+4+2+1=15).

c. The *reader* will ask the bits to count from 0 to 15, one number at a time. The *recorder* will write down the position and placing of each bit. The *reader* should ensure that all team members agree on the positions before moving to the next number.

d. Once all the numbers from 0-15 have been recorded on the counting chart, the *reader* can call out several random numbers (0-15) to the group and challenge them to determine who needs to stand to create the answer.

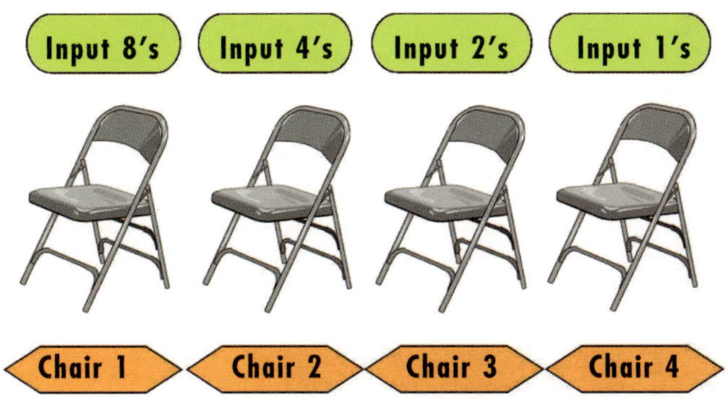

3. Timing is everything — The 4-bit (Nibble) data is collected by the computer and used by the processor. But more data could be coming into the computer. For example, as you type into a computer, it needs to recognize when you are typing different letters on the keyboard. The computer accomplishes this task by collecting the data at specific times or cycles. An internal clock or timer will read the bits or words at each of the time cycles.

Like the previous exercise, four youth will act out the counting process of the binary base 2 system, but this time the status of the inputs will be collected at specific times. Members will use a clock timer to indicate when they should check the positions of the inputs.

A *timer* will start and stop the music like musical chairs. The *bits* can stand up and sit down randomly during the music, but when the music stops, they must remain in their current position (either standing or sitting). A *processor* will record the positions of each bit. Once positions are recorded, the music can start again and the *bits* can randomly stand and sit until the music stops.

a. Rotate the team players so there are new *readers/recorders* and *bits*. This time, they will play a game similar to musical chairs.

b. The new *reader* will be the timer and will start and stop music to indicate when it is time to read the *bits*.

c. While the music is playing, the *bits* will randomly sit and stand at their chairs.

d. When the music stops, the *bits* will stop in their current position to be read.

e. The *recorder* will record the positions for each bit on the timing chart. Once they are recorded, the *timer* will start the music and the *bits* will again randomly stand and sit. Repeat for about 15 cycles.

Sharing and Processing

As the facilitator, help guide youth as they question, share, and compare their observations. Have participants record their experience in their Robotics Notebook, including notes about new discoveries and knowledge. If necessary, use more targeted questions to prompt particular points.

- Ask the teams to compare the completed counting and timing charts to discern any patterns, similarities, or differences.
- How do you think computers and other electronic systems use this binary (base 2) counting?
- How often do you think the computer needs to check for new data? Why is this important? How might this impact a robot?
- What are some advantages and disadvantages of binary versus base 10 counting?

Generalizing and Applying

- Can you think of other places where base 2 counting is or can be used?
- How do you think you could count to 16 in binary? To 32? To 64?
- What are some things that use timers, cycles, pulses, or beats? (metronome, drummers, seasons, days of the week, etc.)
- Extension: Try this activity again with additional chairs added to the left: Chair "16", Chair "32", Chair "64."
- Youth can apply what they have learned in Activity L.

Counting Chart

	Chair 1	Chair 2	Chair 3	Chair 4
Base 10 #	**Base 2 8's**	**Base 2 4's**	**Base 2 2's**	**Base 2 1's**
0	Sit = 0	Sit = 0	Sit = 0	Sit = 0
1				
2				
3				
4				
5	Sit = 0	Stand = 1	Sit = 0	Stand = 1
6				
7				
8				
9				
10				
11				
12				
13				
14				
15	Stand = 1	Stand = 1	Stand = 1	Stand = 1

Timing Chart

	Chair 1	Chair 2	Chair 3	Chair 4	
Clock Cycle #	**Base 2 8's**	**Base 2 4's**	**Base 2 2's**	**Base 2 1's**	**Base 10 # value**
Cycle 0	Sit | 0	Sit | 0	Sit | 0	Sit | 0	0
Cycle 1					
Cycle 2					
Cycle 3					
Cycle 4					
Cycle 5					
Cycle 6					
Cycle 7					
Cycle 8					
Cycle 9					
Cycle 10					
Cycle 11					
Cycle 12					
Cycle 13					
Cycle 14					
Cycle 15					

Activity L – Logic: AND, OR, NOT!

Performance Task For Youth

You will learn about logical operators, including AND, OR, and NOT.

Success Indicators

Youth will be able to demonstrate understanding of common logical operators.

List of Materials Needed

- Robotics Notebook
- Activity Supplies
  - Enough chairs for everyone
  - Copies of logic gate operations cards on pages 57-61
  - String, yarn, or cord in various lengths
- Copies of handout/poster of Logic Gates on page 56

Activity Timeline and Getting Ready

- Activity will take approximately 20 minutes.
- Set up the chairs and string according to the diagram.
- Optional: For durability laminate or glue the logic gate cards on pages 57-61 to card stock. Use a paper punch to make holes near the edge of the paper, one hole for each input and output. In the hole, tie a 3- or 4-foot length of string to represent a conductor.
- Print or copy the handout on Logic Gates so each team has a copy. Use this handout at the end of step 2.
- Divide youth into groups of three or four.

Note to Facilitator

Using the binary system and electronic circuits, a logic system can be applied to help understand input and output controls in a robot. This logic system is a simple binary algebra called Boolean logic. It can be related to electrical circuits and switches, transistors, or integrated circuits using logic gates. Boolean logic has three main logic gates and several combination gates. The logic gates have one, two, or more inputs and a single output. Each of these logic gates has a truth table, which indicates the output results for each of the input combinations. In these activities, youth will explore the main logic gates: AND, OR, and NOT.

Experiencing

1. In teams of four, each youth will be acting as either an input, logical operator, or output. Have youth sit in a chair and read the instructions on their cards. The cards will have a symbol drawing of a logic gate, a truth table for that gate, and strings to connect the inputs and output.

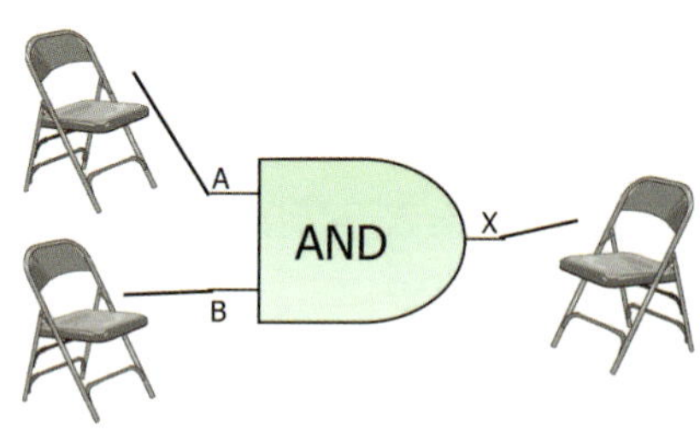

   a. Each member will hold one of the strings from the card. If some members are not holding string, they can read off the truth table.

b. Use the same representation of sitting = 0 (false) and standing = 1 (true). Start by asking the group to use the truth table on their card to determine how the inputs must be set (true or false) so the output is true (1). What input setting(s) would produce an output that is false (0)?

c. Repeat "acting out" the logic with the three main logic gates of AND, OR, and NOT.

2. In the same teams, have an AND or OR gate followed by a NOT gate. The output from the AND or OR holds the string to the input of the NOT gate in his or her other hand. A fourth person holds the output of the NOT gate. How does this change the final output results?

   On a note card, create a new truth table showing the results of the output based upon the new inputs. These combinations are known as NAND and NOR gates. Hand out copies of the NAND and NOR gate cards to confirm their truth tables.

3. Combine two teams and try connecting gates to see different results. For each combination, develop a truth table for the inputs and resulting output.

Sharing and Processing

As the facilitator, help guide youth as they question, share, and compare their observations. Have participants record their experience in their Robotics Notebook, including notes about new discoveries and knowledge. If necessary, use more targeted questions to prompt particular points.

- How many different ways are there to get the desired output?
- What happened if you used three or more input points?
- What are some advantages to using certain logic operators over others?
- Which logic operator was the most useful in this activity?

Generalizing and Applying

- What types of decisions have you had to make? How are the truth tables like a decision maker?
- How do you think these logical operators might be used in the real world?
- Could you build a truth table for how to select which movie you would like to see?
- Youth can apply what they have learned in Activity M.

Logic Gates

The basic logic gates used in binary and electrical circuits include the following symbols and truth tables: Truth tables are listings of all the possible options for the logic in each setting. The table is composed of a column for each input and a column for the output.

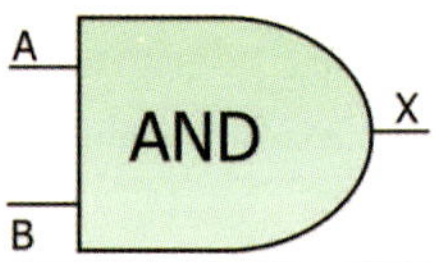

AND truth table		
A	**B**	**X**
0	0	0
0	1	0
1	0	0
1	1	1

Where X = A*B

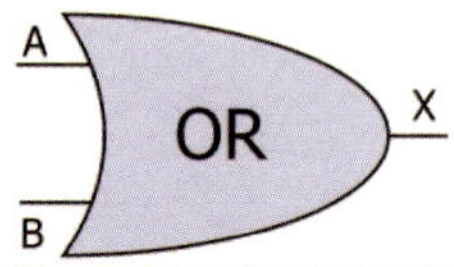

OR truth table		
A	B	X
0	0	0
0	1	1
1	0	1
1	1	1

Where X = A + B

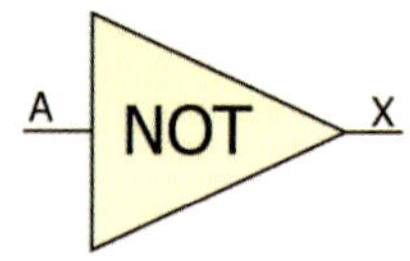

NOT truth table	
A	**X**
0	1
1	0

Where X = not A = B

Combination Gates

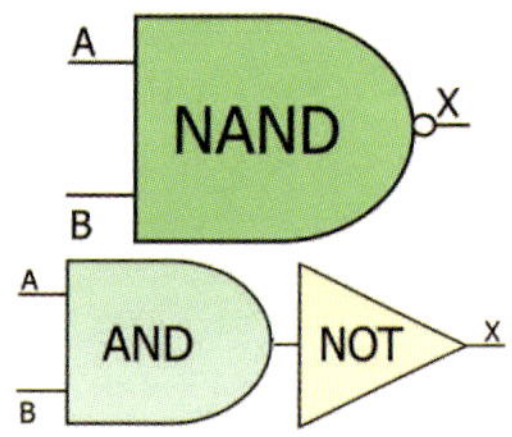

NAND truth table		
A	**B**	**X**
0	0	
0	1	
1	0	
1	1	

Where X = A*B

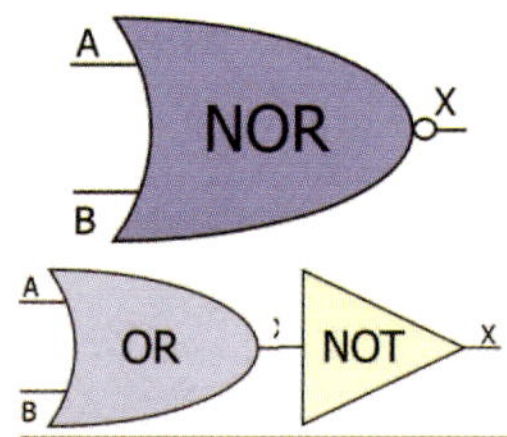

NOR truth table		
A	B	X
0	0	
0	1	
1	0	
1	1	

Where X = A + B

What would the truth tables look like for the NAND and NOR gates?
How are they different from the AND and OR gates?

Acting as an AND Logic Gate

Logic Gate Cards instructions: You may either be TRUE (standing, 1, or on) or FALSE (sitting, 0, or off).

AND truth table		
A	**B**	**X**
0	0	0
0	1	0
1	0	0
1	1	1

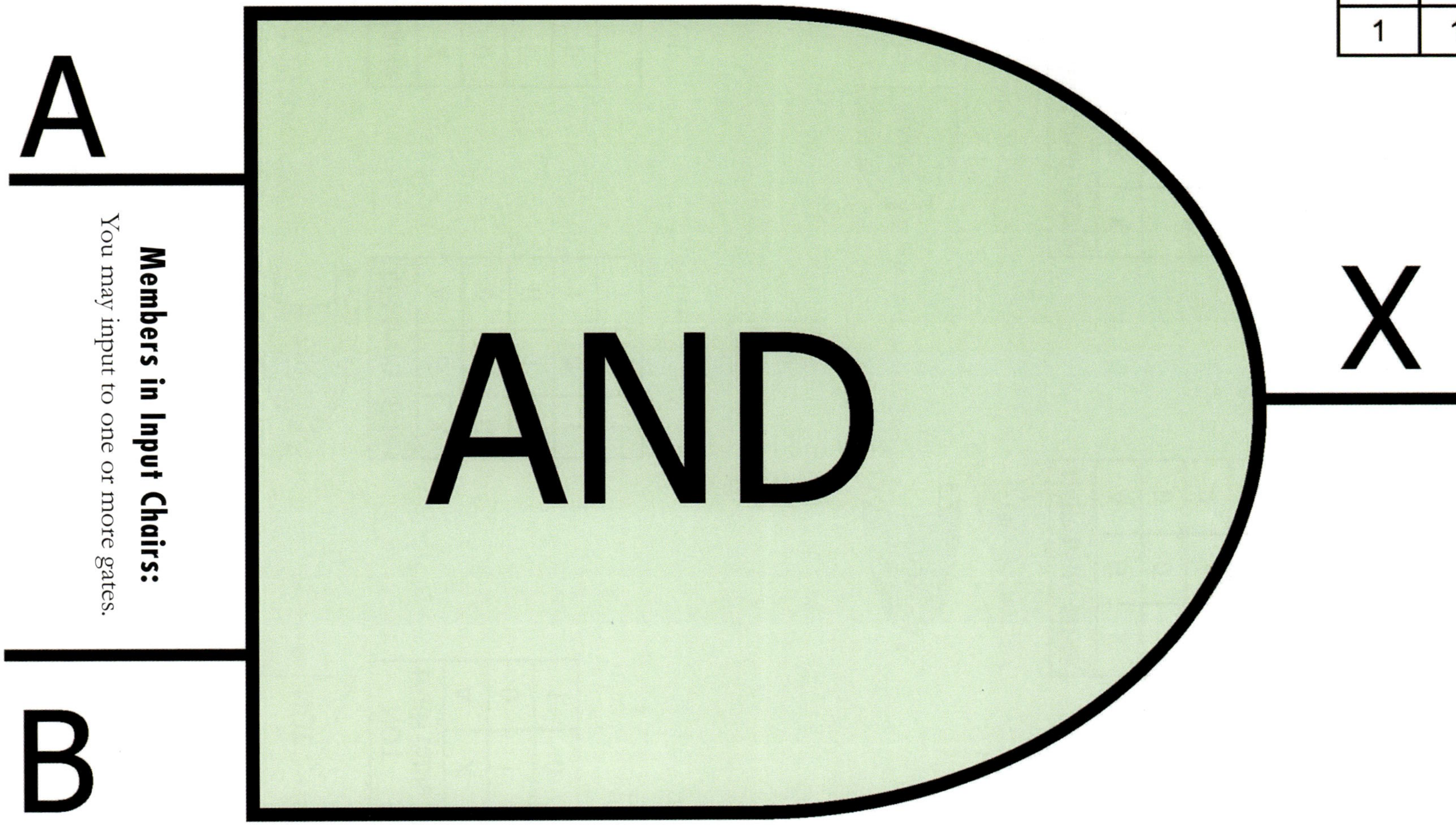

Members in Input Chairs:
You may input to one or more gates.

Member in Output Chair:
You may only receive direction from one gate.
Your state is determined by the truth table.

Logic Gates

The basic logic gates used in binary and electrical circuits include the following symbols and truth tables: Truth tables are listings of all the possible options for the logic in each setting. The table is composed of a column for each input and a column for the output.

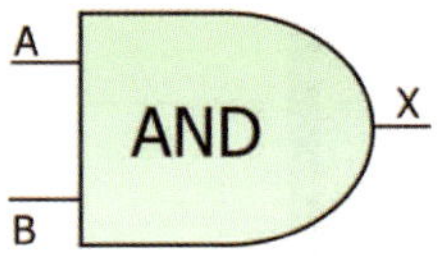

AND truth table		
A	**B**	**X**
0	0	0
0	1	0
1	0	0
1	1	1

Where X = A*B

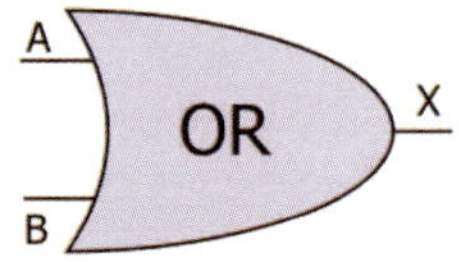

OR truth table		
A	B	X
0	0	0
0	1	1
1	0	1
1	1	1

Where X = A + B

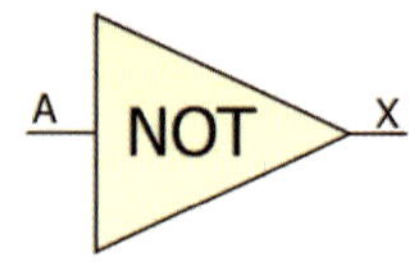

NOT truth table	
A	**X**
0	1
1	0

Where X = not A = B

Combination Gates

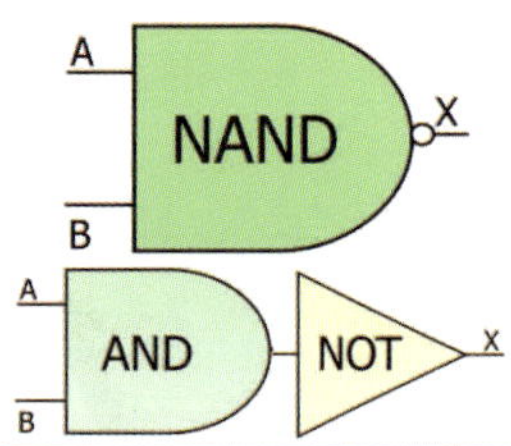

NAND truth table		
A	**B**	**X**
0	0	
0	1	
1	0	
1	1	

Where X = A*B

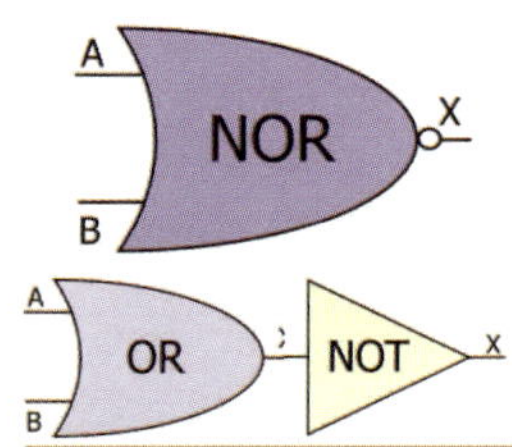

NOR truth table		
A	B	X
0	0	
0	1	
1	0	
1	1	

Where X = A + B

What would the truth tables look like for the NAND and NOR gates?
How are they different from the AND and OR gates?

Acting as an AND Logic Gate

Logic Gate Cards instructions: You may either be TRUE (standing, 1, or on) or FALSE (sitting, 0, or off).

AND truth table		
A	**B**	**X**
0	0	0
0	1	0
1	0	0
1	1	1

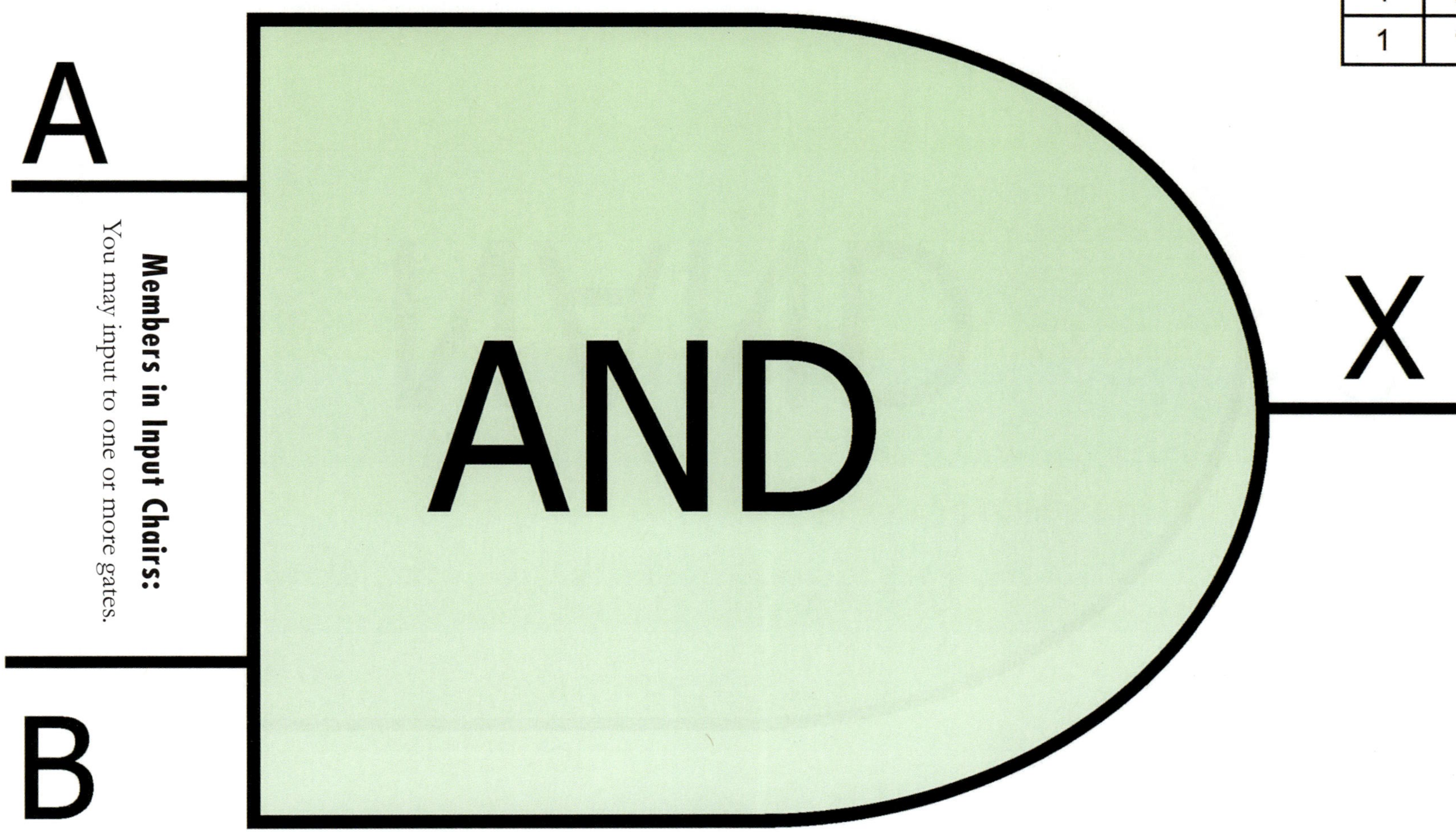

Members in Input Chairs:
You may input to one or more gates.

Member in Output Chair:
You may only receive direction from one gate.
Your state is determined by the truth table.

Acting as an NAND Logic Gate

Logic Gate Cards instructions: You may either be TRUE (standing, 1, or on) or FALSE (sitting, 0, or off).

NAND truth table		
A	**B**	**X**
0	0	
0	1	
1	0	
1	1	

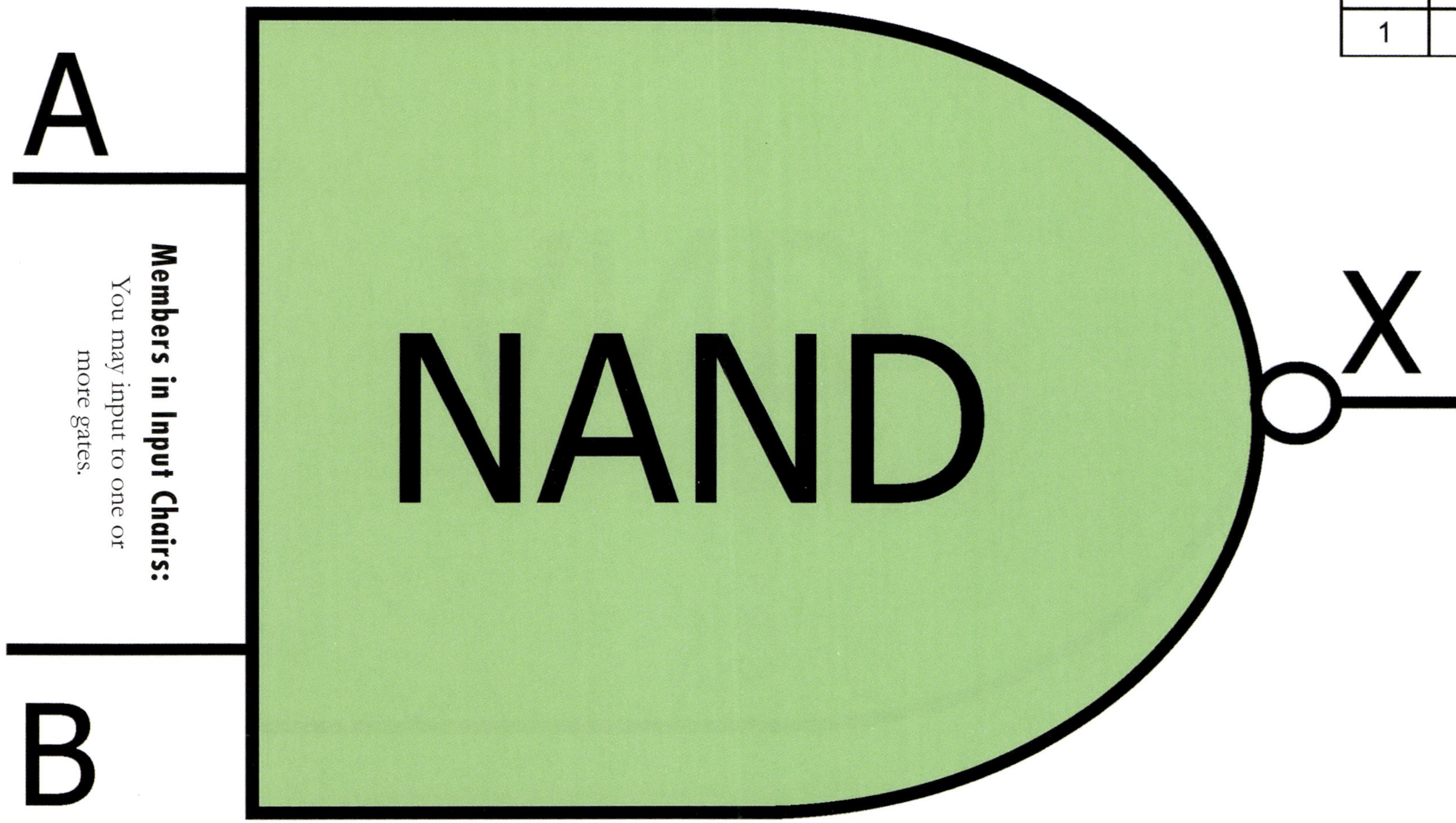

Members in Input Chairs:
You may input to one or more gates.

Member in Output Chair:
You may only receive direction from one gate. Your state is determined by the truth table.

Acting as an NOR Logic Gate

Members in Input Chairs:
You may input to one or more gates.

Member in Output Chair:
You may only receive direction from one gate.
Your state is determined by the truth table.

NOR truth table

A	B	X
0	0	
0	1	
1	0	
1	1	

Activity M – Analog Versus Digital

Performance Task For Youth

You will apply logical circuits in three real-world settings.

Success Indicator

Youth will be able to compare and describe both analog and digital.

List of Materials Needed

- Robotics Notebook
- Activity Supplies
  - Chairs for every youth
  - Copies of logic gate operations cards on pages 57-61
  - Samples of analog and digital interaction situations on pages 64-66
  - String

Activity Timeline and Getting Ready

- Activity will take approximately 20 minutes.
- Divide youth into groups of 10 or more.
- Copy or use logic gate cards from Activity L: inputs, outputs, and logical operators.

Note to Facilitator

Analog is a continuous and variable recording level of many values. For example, temperature can change ever so slightly from cold, cool, warm, hot, and boiling. However, all of these temperature levels would be hard for a computer to handle, so the analog data is converted to digital data. Digital data is specific, as in 1, 2, 3, 4, 5, etc., but there are no "in-between" values. Digital readings are taken at specific levels of analog data and, therefore, share only those analog points. Likewise, digital data can be output as analog by allowing a continuum of values between the specific digital signals. In our activities we will act out some of the inputs and outputs to be analog and some to be digital.

Experiencing

1. Have a discussion with the group about adding together various combinations of logic gates to create circuits that will use inputs to produce specific outputs.
2. Depending on the size of the overall group, have one or more teams use the handouts to set up the following logic circuits using the logic gate cards, members, and chairs. Using the same representation of sitting = 0 (false) and standing = 1 (true), have them enact and verify the truth table for each circuit.
   a. Circuit 1: Temperature Reading (10 members)
   b. Circuit 2: Volume Control (8 members)
   c. Circuit 3: Position Control (12 members)

Sharing and Processing

As the facilitator, help guide youth as they question, share, and compare their observations. Have participants record their experience in their Robotics Notebook, including notes about new discoveries and knowledge. If necessary, use more targeted questions to prompt particular points.

- Where does analog and where does digital work better? Which do you think your body uses?
- What was the easiest part of the activity? The hardest? Why? What do you think the easiest or hardest part would be for a computer?
- How could you build a computer to use analog instead of digital?
- How could the accuracy of the readings be improved? What about the precision? How could it be more sensitive?

Generalizing and Applying

- Where have you seen an analog to digital converter? What other things might use a similar logic circuit to the ones you acted out? What would have to change for it to work?
- Youth can apply what they have learned in Activity N.

Temperature Reading

Ten members act out the following logic that takes an analog reading and comparator inputs to have a digital output. Members sit in a chair and holds string(s) to the gates as shown. One member acts as the analog temperature by walking in front of inputs. As temperature passes input, they go high (1). The temperature can walk back and forth triggering inputs high (1) and low (0).

Inputs – as 0 = sitting and 1 = standing

Gates evaluate their inputs and then show output as 0 = sitting and 1 = standing. A, B are digital **Outputs** shown as 0 = sitting and 1 = standing.

Temperature truth table				
1	**2**	**3**	**A**	**B**
0	0	0	0	0
1	0	0	0	1
1	1	0	1	0
1	1	1	1	1

A= 2
B= (2*3)+not(2+3)+1

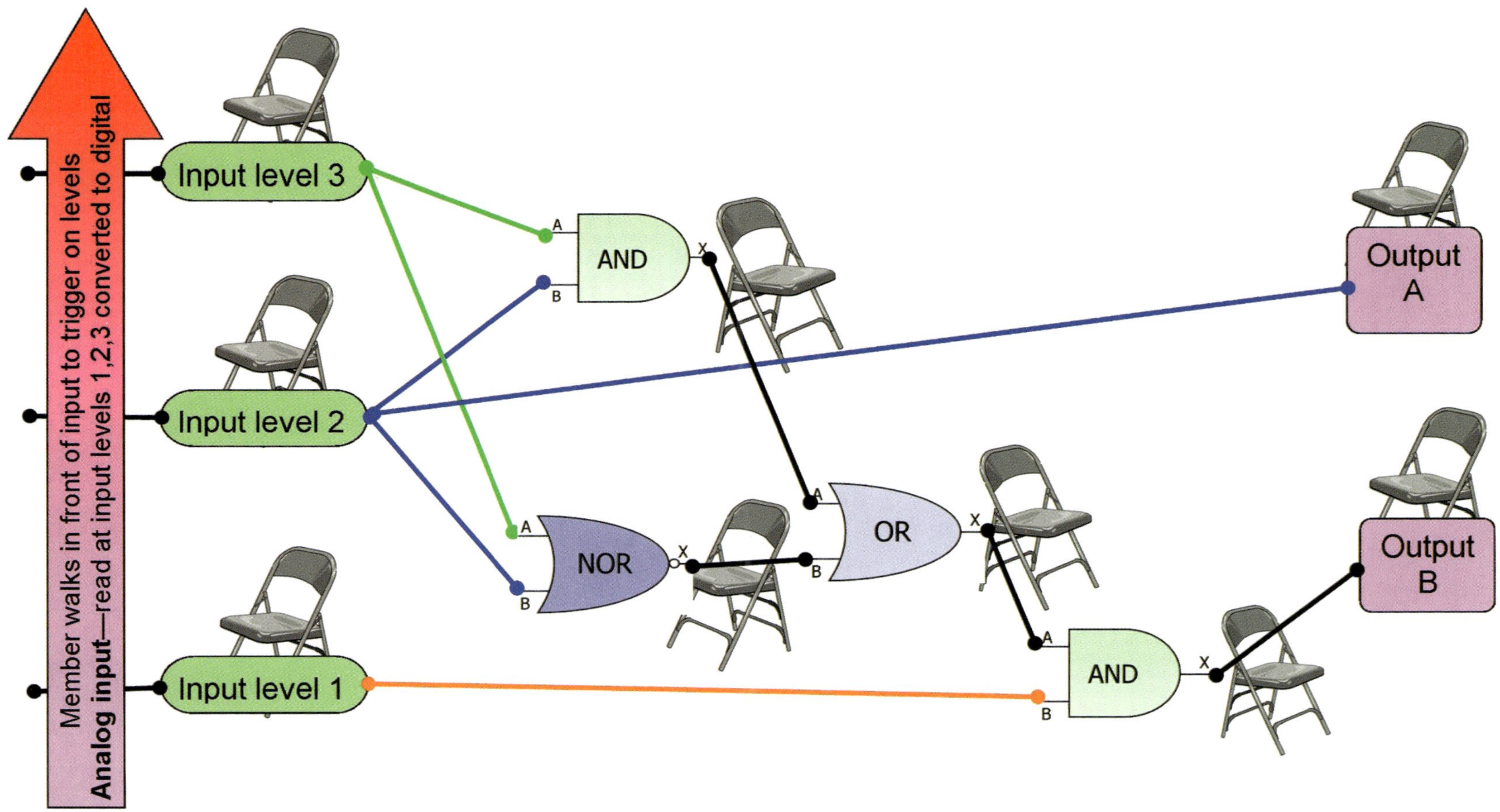

Volume Control

Eight members act out the following logic. Each member sits in a chair and holds a string to the gates as shown.

A, B are digital **inputs** shown as 0 = sitting and 1 = standing.
Gates evaluate their inputs and then show output as 0 = sitting and 1= standing.
Outputs – 0 = sitting quietly and 1 = clapping

Volume truth table				
A	**B**	**1**	**2**	**3**
0	0	0	0	0
0	1	1	0	0
1	0	1	1	0
1	1	1	1	1

1=(AorB)or(AB)
2=A
3=AB

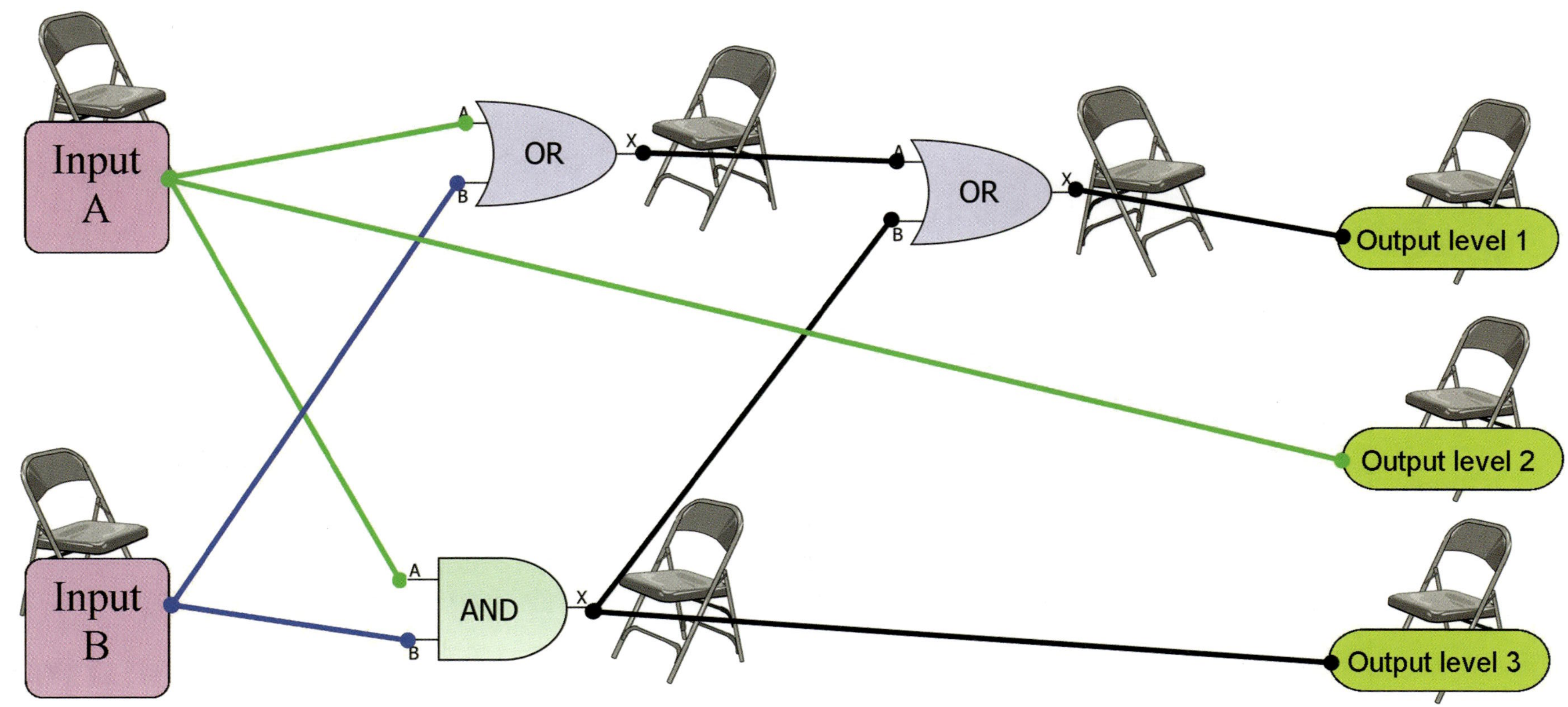

Position Control

Twelve members act out the following logic in which two digital inputs can indicate four different positions. Each member sits in a chair and holds a string to the gates as shown.

A, B are digital **inputs** shown as 0 = sitting and 1 = standing.
Gates evaluate their inputs and then show output as 0 = sitting and 1 = standing.
Outputs – as 0 = sitting and l = standing.

Position truth table

A	B	1	2	3	4
0	0	1	0	0	0
0	1	0	1	0	0
1	0	0	0	1	0
1	1	0	0	0	1

1=notA notB
2=B notA
3=A notB
4=AB

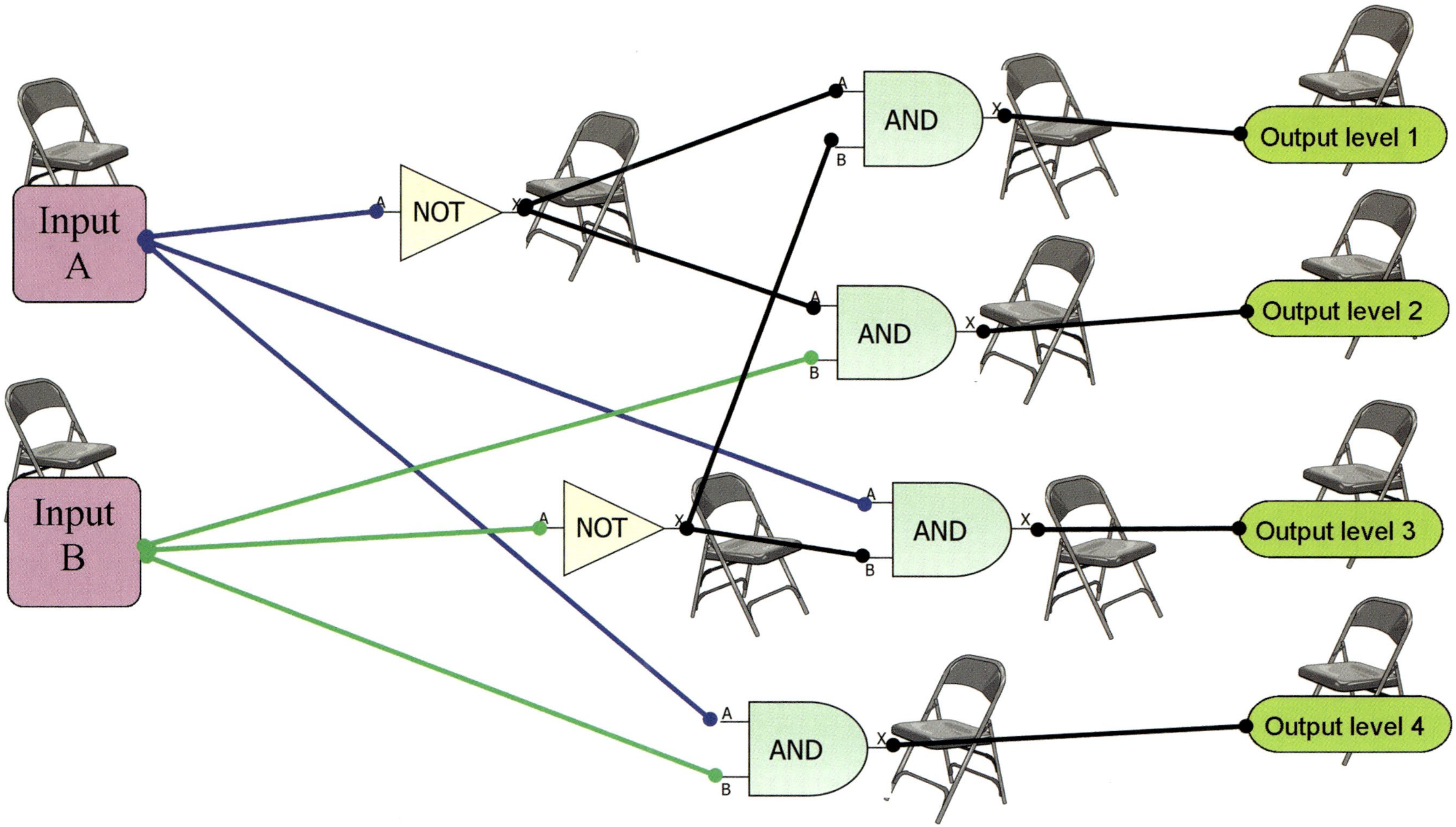

Activity N – Components

Performance Task For Youth

You will learn about electrical components, including batteries, wire conductors, resistors, and capacitors.

Success Indicator

Youth will role play an electrical circuit and act out the roles of batteries, wires, resistors, and capacitors.

List of Materials Needed

- Robotics Notebook
- Activity Supplies
  - Ping-pong or other small nonbreakable balls (20-30 or more)
  - Large container with limited size opening that can hold 20 to 30 ping-pong (or similar) balls
  - Two smaller containers that will only hold about five ping-pong balls
  - Two short pieces of foam pool noodle or pipe insulation
  - Note cards (approximately 5 inches by 8 inches)
  - Clock or watch for time recording
  - A few chairs or small tables
  - An open area to move around in while playing a game
  - Copies of role cards for circuit components on pages 69-71 cut apart, one for each youth

Activity Timeline and Getting Ready

- Activity will take approximately 30 minutes.
- Use groups of at least six, up to the total group of youth. If using more than one group, each group will need its own containers and other supplies.
- Prepare the components from recycled food containers and pool noodle or pipe insulation prior to the activity. See component description list for ideas on how to make the components.
- Optional: Copies of the component descriptions, game instructions, and game circuit schematic on pages 71-73.
- Optional: Lay the game circuit schematic on a table, then place the components on the table in the order shown in the drawing.

Setup

1. Use a table, such as a card table, that all the members can stand around. If the group is large, use a larger table or put two smaller tables together. All of the components will be placed on this table. Make copies of the circuit schematic and component drawing and lay the copy on the table. Then, place the components on the table in the same order as the drawing.
2. Fill the large container (which represents the battery) with all the ping-pong balls and place it on the table. A large, clear plastic, wide-mouth jar with an opening cut near the bottom to release the balls one at a time works well.
3. Place the two smaller containers on the table to represent the C1 capacitor and the LED 1. Next, place the two cut pieces of foam noodle on the table to represent the two resistors.

Game Description for Leader

1. In this game, the members will act as the wire conductors; recycled homemade items will be used to represent the components; and ping-pong balls will represent electrons.
2. Youth may only pass the ball if the receiving member does not currently have a ball. If the next member has a ball, that person should hold the ball until he/she can pass it.
3. Have the members follow the actions described on the role cards for the six indicated positions. If there are more "wire" players or some split roles using parts A and parts B of the role, then adjust the table, space, and action accordingly.
4. When the members get an "electron," they will act on their role, handing off the electron, placing it into a component such as a capacitor, LED, or resistor. They will check their role card to see where they should be getting their

electrons or where they should be giving their electrons.

5. The balls continue being passed until they are returned to the second battery member. The member should toss or drop the ball into the battery jar from about 1 foot away. (Adjust the distance, if necessary, to ensure that some balls make it into the jar and some miss).

Guidelines for Game

1. Players may have only one "electron" in their hand at a time.
2. Players must follow "roles" for their position.
3. Players may only pass one "electron" at a time.
4. Any electrons that are dropped or fall in action are out of play and are to be left on the floor until the game ends.
5. The game continues until there are no moves left for the electrons.
6. Optional starting: Have each player start with one electron in each hand to begin the game.

Experiencing

1. Ask the youth to share with the group what they know about the different types of electrical components used to make an electrical circuit (e.g., batteries, wire conductors, resistors, lights, capacitors, diodes, transistors, and switches). Share with the group items that they did not include in their sharing. These components can be used in the logic circuits that they have been exploring. In most cases, these single components are combined into miniature integrated circuits called ICs or chips. These ICs can process and store memory and conduct many logic actions, and they only require a small space and low voltages.
2. Give role cards to the members in the six positions as they act as the wire conductors for the circuit.
3. Have the members stand around the table and components in the order shown on the schematic. Adjust the spacing of the members and components so that the members can reach their items and the other players.
4. Have the remaining members stand in a line between the components to represent wire or conductor.
5. Give one or two other members a note card (which represents a light bulb) and have them stand between the capacitor and the battery.
6. Have the remaining members stand in a line between the components to represent wire or conductor.
7. Play the game until there are no moves left for the electrons.

Sharing and Processing

As the facilitator, help guide youth as they question, share, and compare their observations. Have participants record their experience in their Robotics Notebook, including notes about new discoveries and knowledge. If necessary, use more targeted questions to prompt particular points.

- In an electrical current, what could the ping-pong balls represent?
- In your own words, describe how individual components work.
- What do the balls that fall on the floor represent?
- What would happen if additional components, such as another battery or resistor, were added to the circuit?

Generalizing and Applying

Describe other objects in the world that have the same effect as these components.

- Thinking about the binary system, timers, and the logic gates, how do these components work together?
- Youth can apply what they have learned in Activity O.

Role Cards for the Components

Game Cards for the six indicated positions. (If there are more players, add in extra "wire" and/or some roles may be split between two players by using parts A and parts B of the role.)

- Players may only have one "Electron" in a hand at a time.
- Players must follow the rules for their position.
- Players may only pass Electrons one at a time.
- Any Electrons that are dropped or fall to the floor are out of play and must be left on the floor until after the end of the game.
- The game continues until there are no moves left for the Electrons.

Player #1

With one hand, take one Electron from the minus "–" side of the Battery. Pass the Electron to your other hand and then offer it to Player #2. You may take another electron from the Battery with your free hand, but once you have both hands full, you have to wait for Player #2 to take the first Electron before getting any more. When there are no more Electrons in the Battery — even if there are more going around in the circuit — lay the Battery on its side.

Player #2

Part A – If Player #1 offers you an Electron, you may use one hand to take it. Pass the electron to your other hand and then offer it to Player #3.

Part B – If Player #1 has no more Electrons from the Battery (laying on its side), check to see if C1 (the Capacitor) has any Electrons stored for use. If so, use one hand to pick up one Electron. Pass the Electron to your other hand and then offer it to Player #3 (or Player #2, if dividing this role).

Player #3

If Player #2 offers you an Electron, you may use one hand to take it. Pass the Electron to your other hand and then check to see if the LED direction of flow will allow you to inset the Electron into the LED. If so, turn the LED to the "on" position. Then check to see if #2 has any more Electrons to offer. When the last Electron is taken and entered into the LED, turn it to the "off" position.

Player #4

Part A – If the LED turns "on" and passes an Electron, you may use one hand to take it. Pass the Electron to your other hand.

Part B – Take the Electron and place it into the R2 Resistor. There can only be one Electron in the Resistor at a time. If there is already an Electron in the Resistor, wait until Player #5 removes it before adding a new Electron.

Player #5

Part A – Check to see if there is an Electron in the R2 Resistor. If so, remove the Electron by squeezing the tube behind the Electron to move it through and out of the other end of the Resistor and place it in your hand.

Part B – Take the Electron and place it into the R1 resistor. Only one Electron can be in the Resistor at a time. If an Electron is already in the Resistor, wait until Player #6 removes it before adding a new Electron to R1.

Part C – If you are waiting to put an Electron into R1, check to see if there is room for another Electron in Capacitor C1. If there is room, toss the Electron into the opening of C1 from about 1 foot away. (Remember, dropped Electrons are lost and out of play.)

Player #6

Part A – Check to see if there is an Electron in the R1 Resistor. If so, remove the Electron by squeezing the tube behind the Electron to move it out of the other end of the Resistor and place it in your hand.

Part B – Take the Electron and toss it into the plus "+" side opening of the Battery. (Remember, Electrons that fall or drop to the floor are lost and out of play.) If the Battery is on its side, it is dead and not rechargeable.

Components

The components in this activity include:

Battery – This is a power source that will cause a flow of electrons (electrical current) when a conductor(s) makes a path to and from the poles (+ and -).

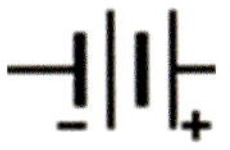

Resistor – A resistor restricts the flow of electrons. Resistors are used to control the flow of electricity in a circuit. The value or amount of resistance in a resistor is identified by a code of bands on the outside of the resistor.

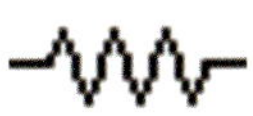

Capacitor – A capacitor is a storage device that can temporarily store an electrical charge to be used for short periods of time.

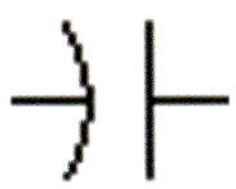

LED – Light Emitting Diode – A diode is a component that will only let electricity flow in one direction. Since electrons can only go through it one way, diodes can control the direction of current in a circuit. LEDs are, as the name implies, diodes that will light up when electricity flows through them.

Wire Conductors – Metals like copper and aluminum make good wire conductors for electricity. They have low resistance and allow electrons to flow freely.

Electrons – In good conductors, electrons can move freely from molecule to molecule when a power source is applied. Insulators, on the other hand, do not allow free movement of electrons and block the flow of electrons or stop the electrical current.

Make the Battery from a cylindrical container that can hold about 15-20 ping-pong balls. Cut a small opening on the top so ping-pong balls (Electrons) can be tossed in the container. Cut an opening near the bottom to let out one ball at a time.

Make the Resistor from a pool noodle or pipe insulator so that a ping-pong ball will barely fit through the center opening. Cut R2 to about 4 inches in length and cut R1 longer, about 6-8 inches in length. Make colored bands around the insulator to mimic the rating code.

Make the Capacitor from a smaller cylindrical container that will only hold about 5-8 ping-pong balls. Make a small opening on the top so ping-pong balls (Electrons) can be tossed in the container. Cut an opening near the bottom to let out one ball at a time.

Make the LED from an old plastic container with a wide mouth lid opening and if possible, one with lift-up flaps such as on some spice containers. Use or make a flap in the lid to allow a ping-pong ball to be dropped into the container (LED). On the opposite side of the flap and near the bottom, cut an opening to allow the ping-pong ball to come out.

For the Wire Conductors, use your arms as the wires and your hands as the molecules that exchange the electrons to create electrical current.

Use ping-pong or other small balls to represent the Electrons. Adjust how you make your Battery, Resistors, LED, and Capacitor components based on the size of the Electrons you will be using.

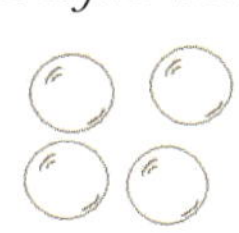

Component Game Instructions

In this game, the members will act as the **Wire Conductors**; recycled homemade items will be the **Components;** and ping-pong balls, the **Electrons**.

Use the schematic and drawing to set up the game and the objects (components). Use the members as the Wire Conductors to connect the components. A minimum of six players is suggested; additional players can be used between some of the basic players to extend the "wire" if you have more members in the class.

Roles for members in the six indicated positions:

- Players may only have one Electron in a hand at a time.
- Players must follow roles for their position.
- Players may only pass one Electron at a time.
- Electrons that are dropped or fall to the floor during action are out of play and must be left on the floor until after the end of the game.
- The game continues until there are no moves left for the Electrons.
- *Optional starting:* Have each player start with one Electron in each hand to begin the game.

Set the components on a table so that the players can stand around the items, reach for, and pass the Electrons.

Components to be made from recycled items, if possible.

Some roles may be split between two players if additional players are used; see parts A and B if splitting roles.

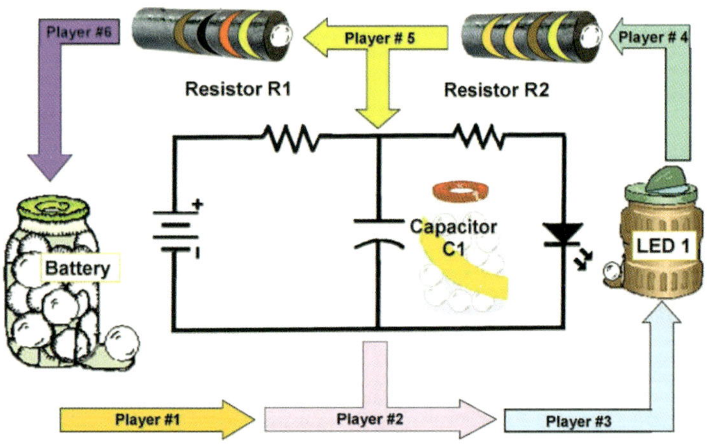

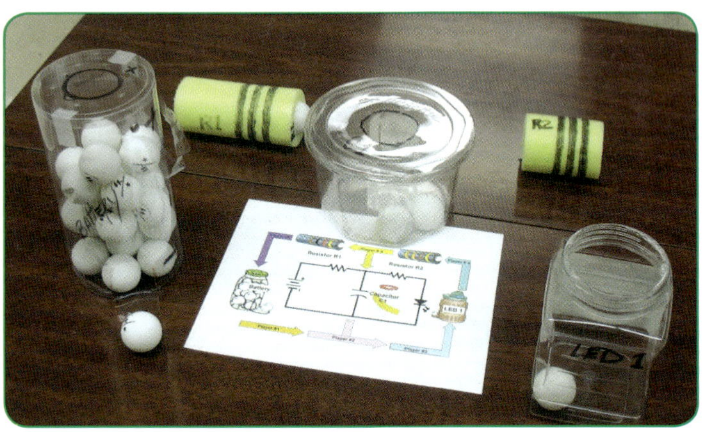

Components

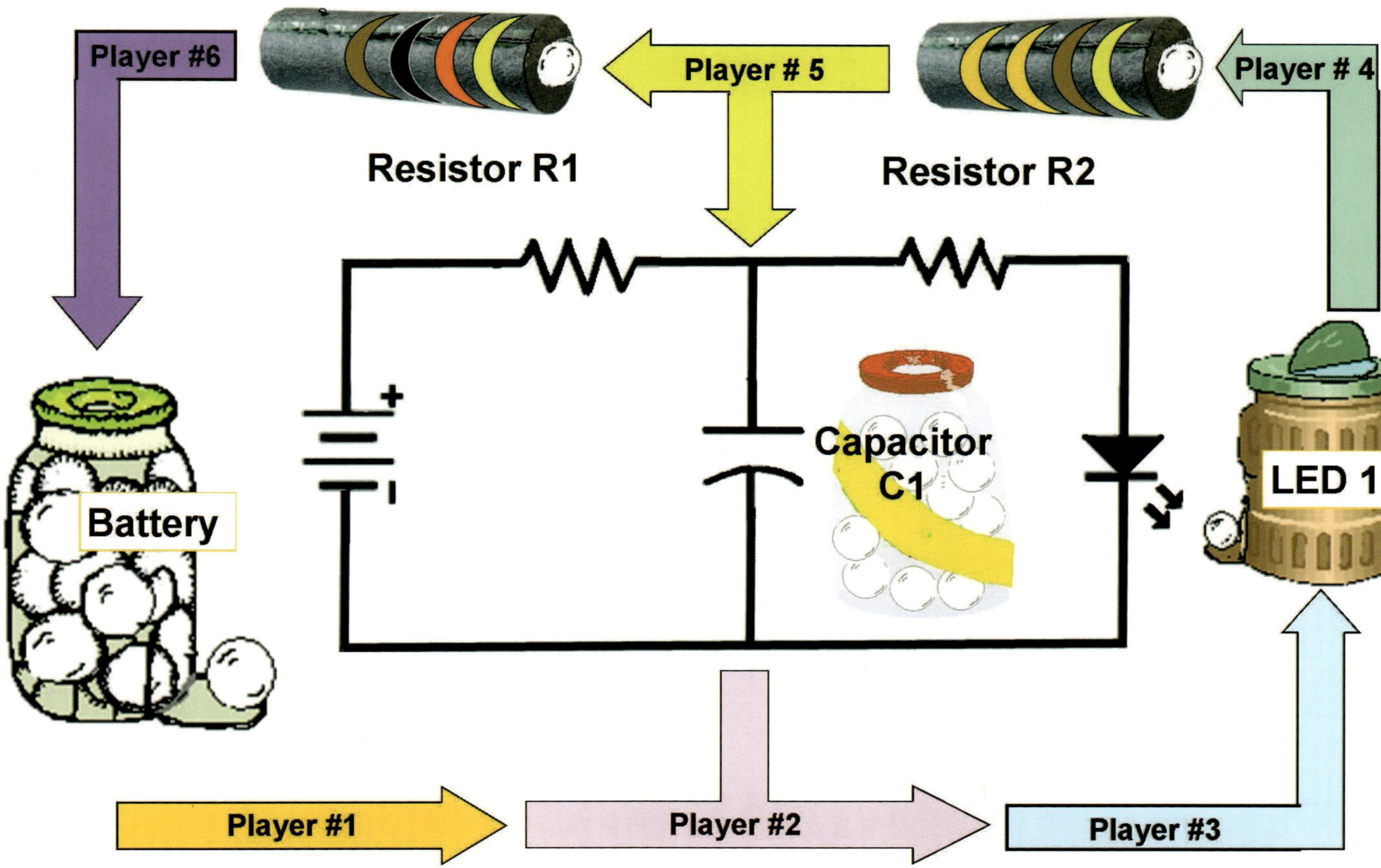
Player #6
Player # 5
Player # 4
Resistor R1
Resistor R2
Battery
Capacitor
C1
LED 1
Player #1
Player #2
Player #3

Activity O – Breadboard Build Team

Performance Task For Youth

Youth will rebuild electronic circuits they've built in previous activities.

Success Indicator

Youth will have created working electronic circuits using a solder-less breadboard.

List of Materials Needed

- Robotics Notebook
- Activity Supplies
  - Electronics Experiment Kit for parts and instructions on these activities, or acquire individual components and develop own experiments using various resistors, LEDs, light bulbs, capacitors, IC chips, etc.
  - Small prototyping breadboard (solder-less breadboard is the best option)
  - 9-volt battery
  - Copies of circuit schematics and instructions from pages 77-79 or those included in your electronics kit

Activity Timeline and Getting Ready

- Activity will take approximately 40 minutes.
- Optional: Sort components required for each of the five circuits on pages 77-79 into small resealable plastic bags and label with circuit number. A few extra items will add to the challenge.
- For larger groups, each team should construct a different circuit.

Experiencing

1. Using information and supplies from Activity N, an Electronics kit, or the resources on the 4-H Robotics website at *www.4-H.org/curriclulum/robotics* and the Junk Drawer Robotics Notebook, lead a discussion about the use of components and a breadboard.
2. Using this knowledge, youth should read the provided schematic diagrams and use components to assemble a basic but functioning electronic circuit. The breadboard activity handout has five different circuits to assemble. The handout shows various elements in circuit design, resistance, capacitance, current control (diode), variable resistance (photocell), and Boolean Logic application.
3. After the teams have constructed and tested their circuit, they should then compare and share with other teams to determine the functions of the five circuits. (See pages 77-79 for more information about the five circuits.)
4. For larger groups, divide into five teams so each team assembles one of the circuit designs. For smaller groups have teams of two to four members.
5. If time allows, the teams may assemble more than one circuit before sharing. Each team will need their own breadboard and components to build their circuit. Note: Resealable plastic bags can be used to sort and store each circuit's components and breadboard. A few extra items in each bag will add to the challenge.

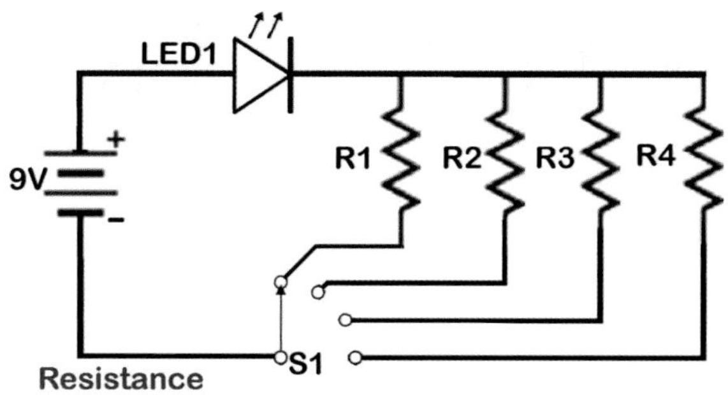

Resistance

- **Circuit 1: Resistance** – Build a circuit that contains a battery source, an LED, and four resistors. The circuit should be wired as shown in the schematic. Allow youth to experiment with connecting resistors of different sizes. In the schematic, switch 1 can just be a wire that has one end that is moved from one resistor contact to the next resistor contact. Make observations on the effect to the LED with each of the different resistors. Have youth record in their Robotics Notebook what happened to the LED when they changed the resistor. Also have youth record the values of each resistor they used in their circuit.

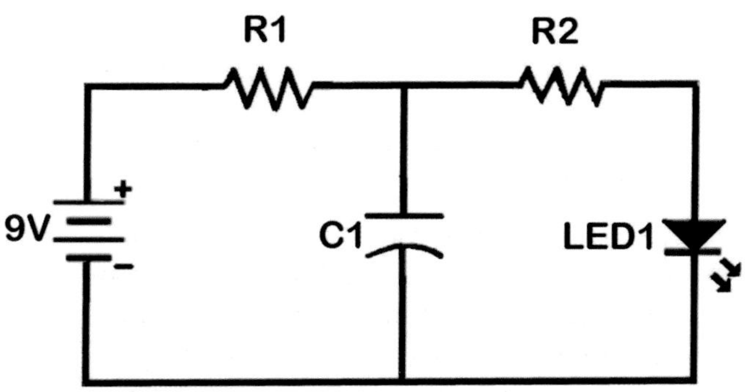

Capacitance

- **Circuit 2: Capacitance –** Build a circuit that contains a battery source, two resistors, a capacitor, and an LED. The circuit should be wired as shown in the schematic. After the circuit is complete, attach the battery to the circuit. Have youth record in their Robotics Notebook what happens to the LED. Then disconnect the battery from the circuit and observe the LED, and again have youth record their findings in the Robotics Notebook. Youth can experiment with capacitors of different sizes (or remove the capacitor) to see how that affects the LED each time the battery is disconnected from the circuit. Then, have youth record in their Robotics Notebook what happened to the LED when the battery was disconnected with each change of capacitor. Also have youth record the values of each resistor and capacitor they used in their circuit.

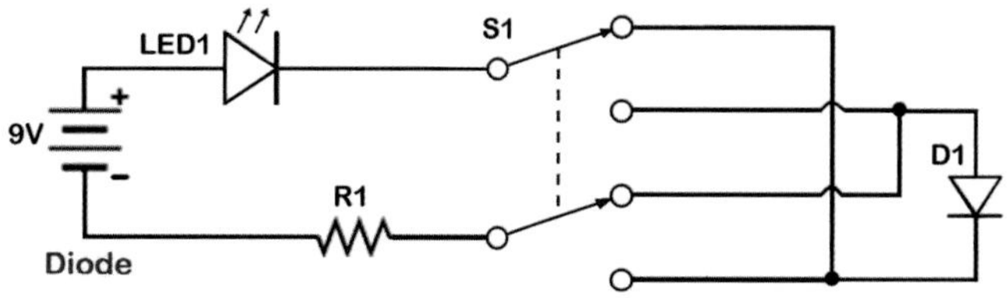

- **Circuit 3: Current Control** – Build a circuit that contains a battery source, resistor, LED, and a diode. The circuit should be wired as shown in the schematic. Have youth change the direction of the diode. In the schematic, switch 1 can just be the two wires with one end of each wire repositioned (switched) to the leads of the diode. Youth should observe the LED and record in their Robotics Notebook the effect of the diode in each switched position in the circuit. Also have the youth record the value of the resistor they used in their circuit.

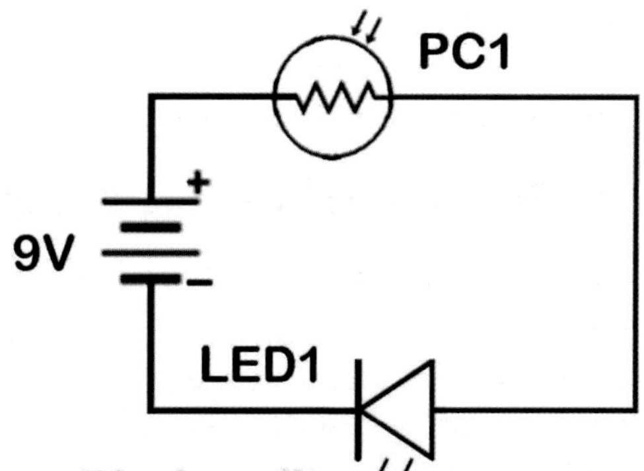

Photocell

- **Circuit 4: Variable Resistance** – Build a circuit that contains a battery source, an LED, and a photocell. The circuit should be wired as shown in the schematic. Have youth experiment with how much light the photocell receives. They can block light with their hand or a piece of paper, or direct more light with a flashlight, mirror, or direct sunlight. Youth should record their observations of the LED in their Robotics Notebook. Ask youth to compare and record the variations caused by the change in light to the results from Circuit 1 on resistance.

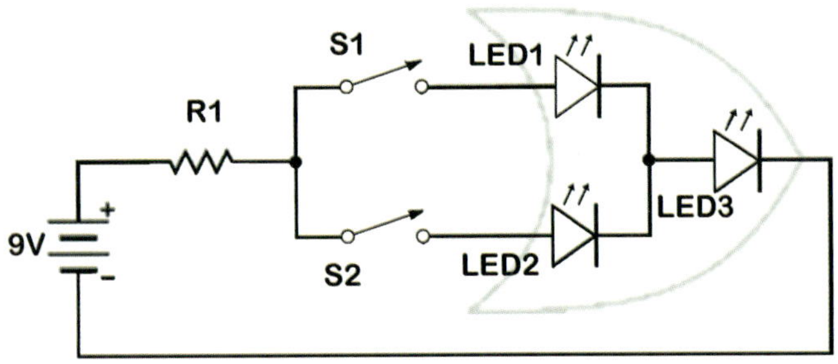

Logic

- **Circuit 5: Logic Circuit** – Build a circuit that contains a battery source, three LEDs, and a resistor. The circuit should be wired as shown in the schematic. In the schematic, switch 1 and switch 2 can be the two wires that each have one end removed to disconnect the flow. Since these LEDs (1 and 2) are in parallel, they can work independently to turn the LEDs off and on. Have the youth observe LED 3 as they turn on and off LED 1 and LED 2. Have youth record their observations in their Robotics Notebook. Also have the youth record the value of the resistor they used in their circuit and build a Truth Table for their Logic Circuit. Have the members determine what type of logic is controlled by their circuit: AND, OR, NOT, NAND, NOR, etc.

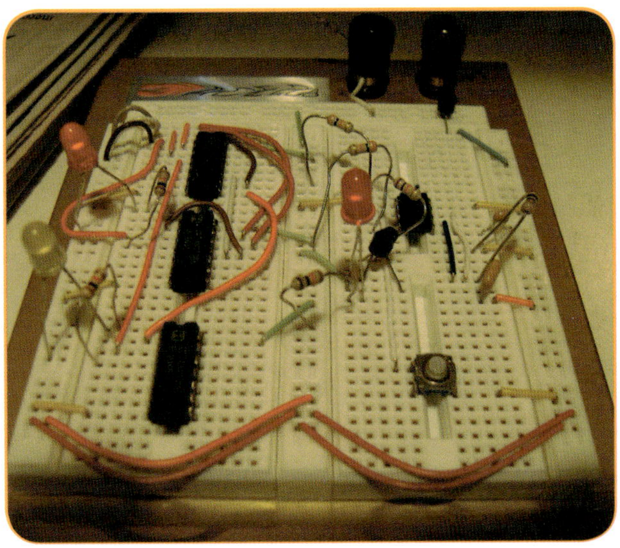

Sharing and Processing

As the facilitator, help guide youth as they question, share, and compare their observations. Have participants record their experience in their Robotics Notebook, including notes about new discoveries and knowledge. If necessary, use more targeted questions to prompt particular points.

- Was this activity more or less challenging than you thought it would be? Why?
- How might one or more of these circuits be used in a robot?
- Which would be the most useful in designing a flashlight?

Generalizing and Applying

Where might some of these circuits be used in the real world?

- Youth can apply what they have learned in the next activity.

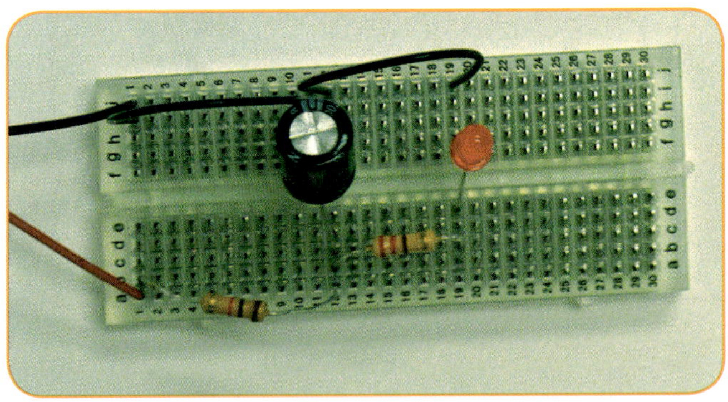

Building Circuits Using Breadboards and Electrical Components

Each team will receive one circuit description, schematic, and instructions to create a sample circuit to highlight how electronic circuits work. Five circuits can be made. Additional sample and more advanced circuits can be made using electronics kits and components for those that want to do more in this area.

Cut each handout page apart so that each team will have notes for their circuit. After the teams have constructed and tested their circuit, they should compare and share with other teams to determine the functions of each of the five circuits.

Notes:

- LED of other colors besides red can be used.
- Resistor ranges are given to assist in selecting, but also to allow variations. Any resistor value in the range should work.
- In circuit 1 on resistance, the only concern is to have enough value difference between resistors so that the LED will be able to show a light intensity change.

Resources:

- Component Game Handout from Activity N
- The 4-H Robotics website, *www.4-H.org/curriculum/robotics*, has resources on how to use a breadboard and electronic components.
- Other books or kits for electronics

Resistance

Components

Power source - 9-volt battery
LED1 red LED
Resistors (range to select from)
R1 100-300 ohmΩ
R2 500-1000 ohm
R3 1500-2000 ohm
R4 12,000-16,000 ohm (16KΩ)
Note:
In selecting resistors allow enough value difference between R1, R2, etc.

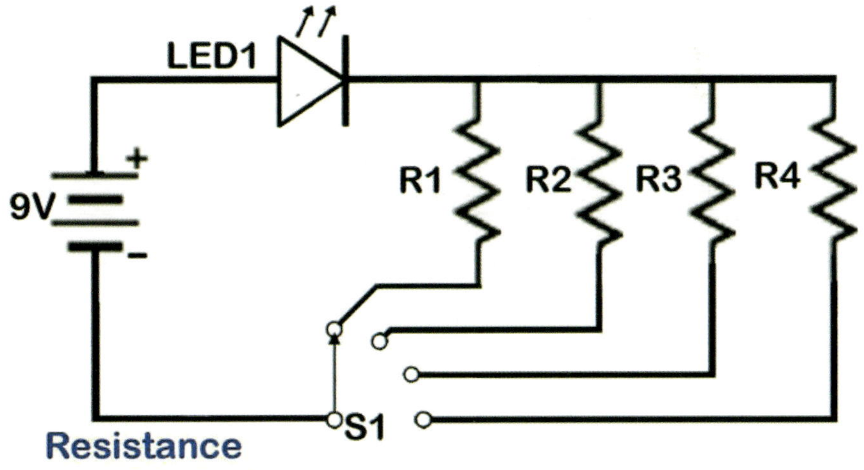

Circuit 1: Resistance – Build a circuit that contains a battery source, an LED, and four resistors. The circuit should be wired as shown in the schematic. Experiment with connecting the different value size resistors. In the schematic, switch 1 can just be a wire with one end that is moved from one resistor contact to the next resistor contact. Make observations on the effect to the LED with each of the different resistors. Then record in your Robotics Notebook what happened to the LED when you changed the resistor. Also record the values of each resistor you used in your circuit.

Capacitance

Components

Power source -	9-volt battery
Resistors (range to select from)	
R1	1,000-1,200 ohm
R2	200-500 ohm
LED1	red LED
Capacitor (µF = micro farad)	
C1	100 µF to 1,000 µF

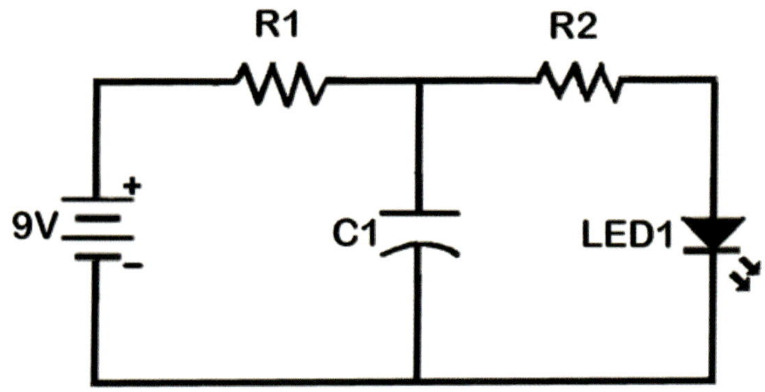

Capacitance

Circuit 2: Capacitance – Build a circuit that contains a battery source, two resistors, a capacitor, and an LED. The circuit should be wired as shown in the schematic. After the circuit is complete, attach the battery to the circuit. Record in your Robotics Notebook what happens to the LED. Then disconnect the battery from the circuit and observe the LED, again record findings in the Robotics Notebook. You can experiment with different size capacitors (or remove the capacitor) to see how it affects the LED each time the battery is disconnected from the circuit. Then record in your Robotics Notebook what happened to the LED when the battery was disconnected with each change of capacitor size. Also record the values of each resistor and capacitor you used in your circuit.

Current Control

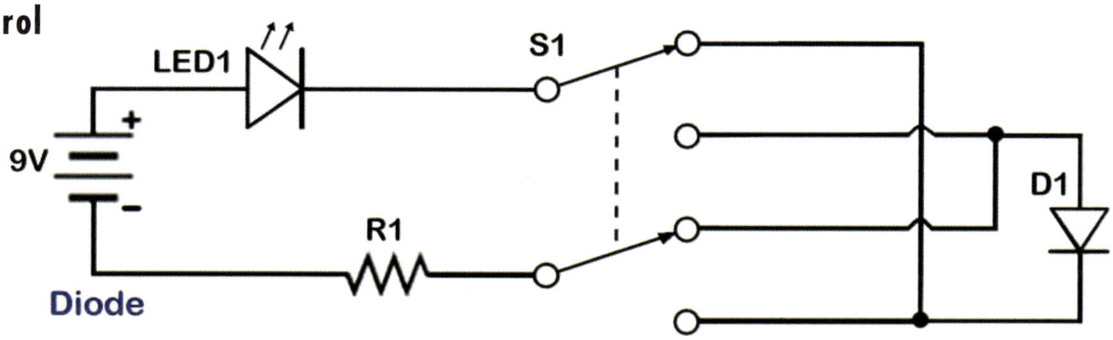

Diode

Components

Power source -	9-volt battery
LED1	red LED
Switch1	DPDT (switch wires)
Diodes	
D1	diode
Resistors (range to select from)	
R1	200-500 ohm

Circuit 3: Current Control – Build a circuit that contains a battery source, a resistor, an LED, and a diode. The circuit should be wired as shown in the schematic. In the schematic, switch 1 can just be the two wires repositioned so that one end of each wire is switched to the leads of the diode. Try the diode one way and record, then change the direction of the diode (switch the wires or the diode). You should observe the LED and record in your Robotics Notebook the effect of the diode in each switched position in the circuit. Also record the value of the resistor you used in your circuit.

Variable Resistance

Components

Power source -	9-volt battery
Photocell	light sensitive cell
LED1	red LED

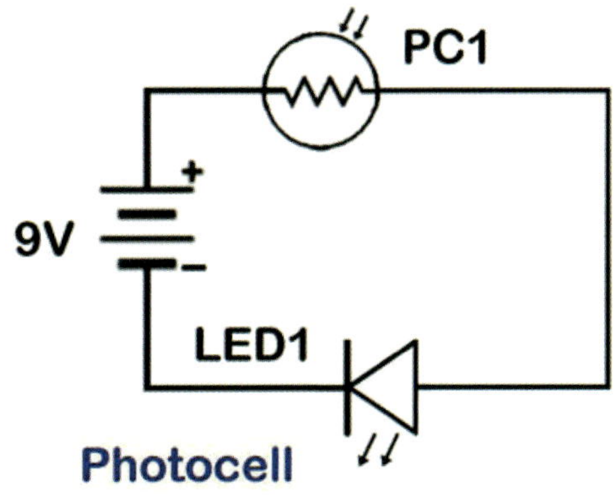

Circuit 4: Variable Resistance – Build a circuit that contains a battery source, an LED, and a photocell. The circuit should be wired as shown in the schematic. Experiment with how much light the photocell receives. You can block light with your hand or a piece of paper, or direct more light with a flashlight, mirror, or direct sunlight. You should record your observations of the LED in your Robotics Notebook. Compare and record the variations caused by the change in light to the results from Circuit 1 on resistance.

Logic Circuit

Components

Power source -	9-volt battery	
Resistors (range to select from)		
R1	200-500 ohm	
Switch1	SPST (move wire)	
Swith1	SPST (move wire)	
Light Emitting Diodes		
LED1	red LED	(A)
LED2	red LED	(B)
LED3	red LED	(X)

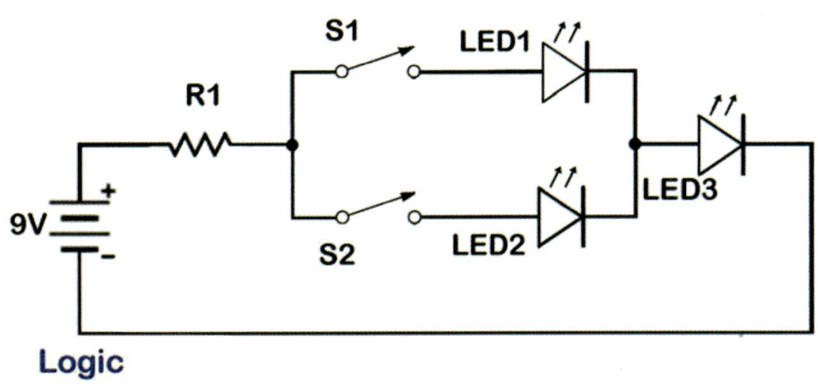

Circuit 5: Logic Circuit – Build a circuit that contains a battery source, three LEDs, and a resistor. The circuit should be wired as shown in the schematic. In the schematic, switch 1 and switch 2 can be the two wires that each have one end removed to disconnect the flow. Since these LEDs (1 & 2) are in parallel, they can work independently to turn the LEDs off and on. Observe LED 3 as you turn on and off LED 1 and LED 2. Record your observations in your Robotics Notebook. Also record the value of the resistor you used in your circuit and build a Truth Table for the Logic Circuit. Determine what type of logic is controlled by your circuit, AND, OR, NOT, NAND, NOR, etc.

CAREER CONNECTIONS

Engineering Professions

Engineers develop solutions to problems in every aspect of life. However, it would be impractical for each engineer to know a lot about everything, so engineers specialize in a field of engineering, such as electricity, civil, medicine, machines, chemistry, or robotics.

There are almost 300,000 electrical engineers in the United States designing electronic equipment. They are responsible for a wide range of technology and often specialize in a specific area, like robot control systems. Electronic engineers provide technologies such as GPS and automobile wiring.

Civil engineering is considered one of the oldest branches of engineering. The basic duties of civil engineers are design and construction of bridges, roads, buildings, dams, and other structures. While some civil engineers manage large scale projects, others do research on improving materials or designs.

Biomedical engineers combine engineering with biological and medical concepts. They develop devices and procedures that solve health-related problems, such as artificial organs. Many biomedical engineers do research on the relation of engineering and the human body, finding robotic replacements for limbs.

Industrial engineers determine the most effective ways for companies to use basic components like people, machines, and information. They are interested in increasing efficiency and productivity, so they devise management and production systems.

Mechanical engineers design and manufacture machines, engines, tools, and other mechanical devices. They develop machines that produce power, such as generators and turbines, and machines that use power, such as air conditioners and elevators.

Although these engineering professions focus on different fields, they share many commonalities. All engineers follow a similar design process. They use notebooks to record their findings, and they test and retest their devices. Ultimately, all engineers use critical thinking to devise solutions for everyday problems. Engineers in different fields work together and combine their knowledge to create the best possible outcomes.

- Which engineering career best fits your interests?
- Which engineering fields are involved in designing and building robots?

Module 4: Do What I Say!

Overview of Activities in this Module

To Learn

Activity P – Cashier
Activity Q – Walk the Walk

To Do

Activity R – Say What? Design Team

To Make

Activity S – Say What? Build Team

Note to Leader

How does a robot recognize its location? How does it know which function to perform at a given moment? Plant, animal, human, and robotic systems around us react to environmental changes based on feedback from their sensors. In a robotic system, the information from these sensors triggers programs that are designed to perform specific functions. Similarly, many robots use a computer program to assist them in their work. Just as a store door opens when you arrive, the robot uses programs with inputs, processing, and outputs.

A computer program is a list of steps or instructions. These instructions may be written in words, statements, or codes that are translated into a binary form for the computer to use and act on. The robot's computer uses a program to read (input), manipulate data (process), and provide or perform action (output). In a circular fashion, the programs instruct the robot, and the robot gives feedback to the program.

To develop or design a program, the first task is to determine what we want the robot to do and how to do it. Once the robot task is selected, a flowchart can be created. A flowchart is used to describe the order and steps to be taken through symbols. These graphics show input and output, decisions, splits or forks, and subroutines.

After the flowchart and design process is complete, the instructions can be written. This is the writing code of the steps or instructions for the computer to use in processing inputs. The instructions have to be written in a language that the computer can read, understand, and follow.

Just as there are many human languages, there are many different computer languages. Some of the most utilized languages include Java, C, C++, C#, and PHP. This module will explore how to develop a flowchart that could be used with any of the many computer languages.

If the robot acts differently than wanted or expected, it may have been given poor instructions. Testing, changing, and rechecking to ensure correct instructions to produce the desired outcome is called debugging.

Once a robot program is completed and ready to use, a set of human instructions can be useful for operating the program and the robot. This written manual or documentation allows others to understand, control, and use the robot.

The following steps will be found in most programs, as shown in the following example. The process of opening an automatic door can be expressed in three steps:

1. Input: The sensor detects someone is coming and tells the program.
2. Program: The program "hears" from the sensor that someone is present and sends power to the motors in the doors.
3. Output: The motors open the doors.

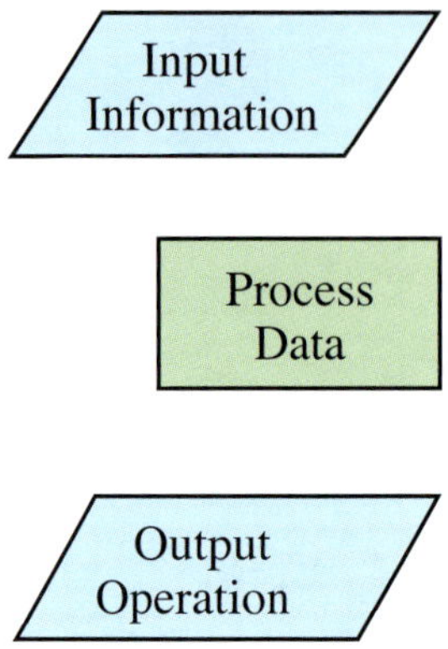

What You Will Need for Module 4: Do What I Say!

- Robotics Notebook for each youth
- Activity Supplies
  - Copies of instructions on page 85 cut apart, one for each youth
  - Copies of bar code samples on pages 86-89 cut apart and attached to notecards
  - Copies of Handout 3 on page 90 for each person playing the "processor" role
  - Optional: Make a handout/poster of the Flowchart Elements on page 94
  - Copies of Flowchart on page 95 or from the Robotics Notebook for each person playing the "processor" role
  - Copies of instructions on page 97 for each participant
  - Copies of the Walk the Walk Memory Cards on page 98 for each person playing the "data memory" role
  - Note cards
  - Tape
  - Graph paper
  - Floor space
  - Blank layout grids
  - Barcodes with unique numbers (10)
  - Index of all the unique barcode numbers and the price list with corresponding products and prices
  - Whiteboard, chalkboard, poster paper, or notebook

Timeline for Module 4: Do What I Say!

Activity P – Cashier

- Activity will take approximately 20 minutes.
- Break the group into teams of four to act out the roles in the programs.
- Make copies of the game instructions and barcodes, and place them on note cards. Provide poster board, clipboard, notebook, or paper for recording results.

Activity Q – Walk the Walk

- Activity will take approximately 20 minutes.
- Divide youth into teams of four.
- Using the layout guide, indicate or mark a grid on the floor.

Activity R – Say What? Design Team

- Activity will take approximately 20 minutes.
- Divide youth into teams of four.
- Obtain all necessary materials. Participants should have an area to write on.

Activity S – Say What? Build Team

- Activity will take approximately 20 minutes.
- Use the same groups from Activity R, Say What? Design Team.

Focus for Module 4: Do What I Say!

Big Ideas

- Programming allows us to control robot behavior.
- Basic logic elements in programming allow us to predict outcomes.
- Flowcharts are often used in programming to help clarify instructions.
- Input, output, if-then-else, loops, goto, and other commands are core components of simple programs.

NSE Standards

- Constancy, change, and measurement
- Abilities of technological design
- Understanding about science and technology

STL

- Relationships and connections
- Attributes of design
- Problem solving
- Information and communication technologies

SET Abilities

- Decision Making
- Sequencing
- Communicate/Demonstrate

Life Skills

- Keeping records
- Self-Discipline
- Teamwork

Performance Tasks For Youth

You will illustrate the concepts of input, processes, and output through an activity simulation.

You will follow simple flowchart instructions to act out the role of processor in a simple computer program.

You will plan and design a program using the flowchart format involving basic commands like loops, goto, and if-then-else so the program can be acted out on a grid.

You will follow the program designed by another group and offer suggestions for improvement.

Success Indicators

Youth will be able to describe the concepts of input, process, and output.

Youth will be able to follow and understand flowchart processes, including if-then-else and goto in a simple program.

Youth will create a flowchart program that includes standard symbols and actions to act out if-then-else, goto, and loops.

Youth will be able to review, follow, and make improvements to another team's flowchart program.

Activity P - Cashier

Performance Task For Youth

You will illustrate the concepts of input, processes, and output through an activity simulation.

Success Indicators

Youth will be able to describe the concepts of input, process, and output.

List of Materials Needed

- Robotics Notebook
- Activity Supplies
  - Copies of instructions on page 85 cut apart, one for each youth
  - Copies of bar code samples on pages 86-89 cut apart and attached to notecards
  - Copies of Handout 3 on page 90 for each person playing the "processor" role
  - Note cards
  - Tape
  - Floor space
  - Barcodes with unique numbers (10)
  - Index of all the unique barcode numbers and the price list with corresponding products and prices
  - Whiteboard, chalkboard, or poster paper, if not using Robotics Notebook

Activity Timeline and Getting Ready

- Activity will take approximately 20 minutes.
- Break the group into teams of four to act out the roles in the programs.
- Make copies of the game instructions and barcodes, and place them on note cards. Provide poster board, clipboard, notebook, or paper for recording results.

Note: This activity is a modification of an activity idea by James Isom provided for an earlier version of this curriculum.

Experiencing

1. In teams of four, members should act out the cash register process. Assign each member a role to play – *Shopper, Input, Processor,* or *Output.*
2. Give each player the instructions on the note cards.
3. Have them stand in a single file line with the *Shopper* in the front, followed by the *Input*; then the *Processor*, and then the *Output* closest to the whiteboard, clipboard, or notebook.
4. As the members perform their role, have them turn around and pass their information to the next person in line.

Cashier Handout 1

Instructions: Copy, cut, and tape on note cards or other individual sheets.

Instructions for the *Shopper*:

Turn around and show one of the barcode pages to the *Input*, then turn around and wait until the *Output* says, "BEEP" or "ERRR" to show the next card. Show some cards quickly, some slowly. Obscure some numbers to ensure that your group will have some errors, but you want to create mostly good reading results.

Instructions for the *Input*:

As a barcode is shown to you, note the barcode number on the recording chart. If the number passes too quickly to read, just pass on what you remember. Then turn to the *Processor* and say the number. Say the code only once, then turn back to your original position and wait for the next barcode to be shown.

Instructions for the *Processor*:

Wait for the *Input* to share a barcode number. Look it up on the product index list. Then turn to the *Output* and say the item name and price. If the number was not on the product list, tell the *Output* that the product doesn't exist. Then turn back around and wait for the next product number from the *Input*.

Instructions for the *Output*:

Wait for the *Processor* to turn to you and give a product name and price. Write it down on your chart or poster. After you write the product and the price, say "BEEP!" If the *Processor* tells you the product doesn't exist, write "N/A" and say "ERRR!"

Cashier Handout 2

Bar code samples – copy, cut, and tape on note cards or other individual sheets.

0 0 0 0

0 0 0 1

0 0 1 0

1 0 1 0

1 0 1 1

1 1 0 0

0 1 1 0

1 1 1 1

1 1 1 0

0 1 0 1

1 0 1 1

0 0 1 0

Cashier Handout 3

Junk Drawer Robotics – Cashier						
Barcodes				**Products**		**Price**
0	0	0	1	Eggs	=	$1.75
0	0	1	0	Bread	=	$2.00
0	0	1	1	Cake	=	$4.25
0	1	0	0	Apples	=	$3.10
0	1	0	1	Milk	=	$2.50
0	1	1	0	Hot Dogs	=	$1.75
0	1	1	1	Bananas	=	$1.50
1	1	0	0	Meat	=	$5.25
1	1	0	1	Juice	=	$3.10
1	1	1	0	Rolls	=	$2.00
1	1	1	1	Cereal	=	$2.75

Sharing and Processing

As the facilitator, help guide youth as they question, share, and compare their observations. Have participants record their experience in their Robotics Notebook, including notes about new discoveries and knowledge. If necessary, use more targeted questions to prompt particular points.

- Can any of the three roles be broken down further?
- What shortcuts can be made in the program?
- How are our senses and sensors like a robot?
- What are some examples of "ouch" reflexes? (e.g., pulling your hand away from a flame, seeing a fire, and calling 911) See if you can find other examples of this. Identify the inputs, programs, and outputs for each.

Generalizing and Applying

- What "control systems" do humans have?
- How are humans programmed?
- How can new features like a running total or multiples of the same item be added?
- Youth can apply what they have learned in Activity Q.

Activity Q - Walk the Walk

Performance Task For Youth

You will follow simple flowchart instructions to act out the role of processor in a simple computer program.

Success Indicators

Youth will be able to follow and understand flowchart processes, including if-then-else and goto in a simple program.

List of Materials Needed

- Robotics Notebook
- Activity Supplies
  - Optional: Copies of Flowchart on page 95 for each person playing the "processor" role
  - Copies of instructions on page 97 for each participant
  - Copies of the Walk the Walk Memory Cards on page 98 for each person playing the "data memory" role
  - Tape
  - Graph paper
  - Floor space
  - Toy blocks, ping-pong balls, etc.

Activity Timeline and Getting Ready

- Activity will take approximately 20 minutes.
- Divide youth into teams of four.
- Using the layout guide, indicate or mark a grid on the floor.

Experiencing

1. Using a grid on the floor (tile squares, grid pattern, or a grid marked with masking tape), walk the grid using the instructions on the grid handout. Use the Flowchart Elements handout to teach the youth different commands.
2. Use the Walk the Walk handout to help teams complete the program task.
3. Directions for performing the Walk the Walk program and the players' roles:
   a. Locate or mark a grid on the floor to fit the pattern shown in the Walk the Walk Grid Setup handout.
   b. Place a chair, box, or basket in the squares marked "pick up blocks" and "place blocks."
   c. In the pick up box, place a number of items (toy blocks, ping-pong balls, etc.) to be picked up and carried to the place box.
4. Place participants into teams of three or four people.
5. Assign roles:
   a. **Actuator** – motion control and movement (input and output)
   b. **Processor** – reading instructions, making decisions, giving feedback
   c. **Data memory** (one or two members per team) – doing math and keeping count of the number of steps and blocks during the activity.

6. The first person will be the *actuator* and will represent the movement of the robot. The *actuator* will stand on the start/end square.
7. The second person will be the *processor* and will read the flowchart program. The *processor* will give commands to the *actuator* and give/receive data from the *data memory*. He or she will stand on the *processor* square.
8. The third person, and fourth person if teams have four members, will play the role of *data memory* and will store variable data input/output on a data memory note card as directed by the *processor*.
   a. If the team has three members, the data memory will store both the steps and block data.
   b. If the team has four members, one person will store the step data and the other will store the block data.
9. The *processor* will begin by reading the flowchart program, one box at a time. See the sample in the handouts.

Sharing and Processing

As the facilitator, help guide youth as they question, share, and compare their observations. Have participants record their experience in their Robotics Notebook, including notes about new discoveries and knowledge. If necessary, use more targeted questions to prompt particular points.

- Ask each participant to share how he or she followed the instructions.
- How did the different players interact? (actuator, memory, processer)
- How might the roles be combined or split?

Generalizing and Applying

How could you change the flowchart to pick up more blocks or balls?

- How would you add sensors and instructions to pick up only square blocks or a specific size ball?
- Youth can apply what they have learned in Activity R.

Flowchart Elements

Terminal – used to show an action, such as start or stop, in a program.

Input/Output – represents the acquisition of information or the telling of data.

Process – used to represent a general process, like a change in value, location, or form of data.

Decision – used to show how choices can cause branches or loops as alternate paths.

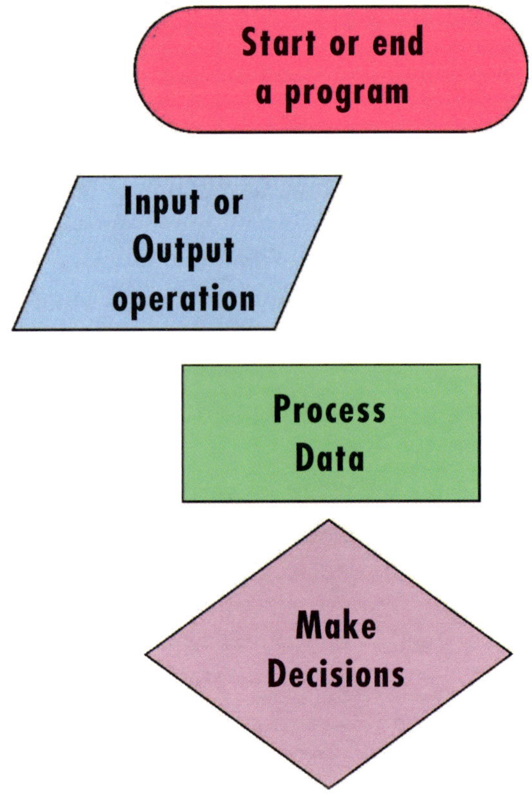

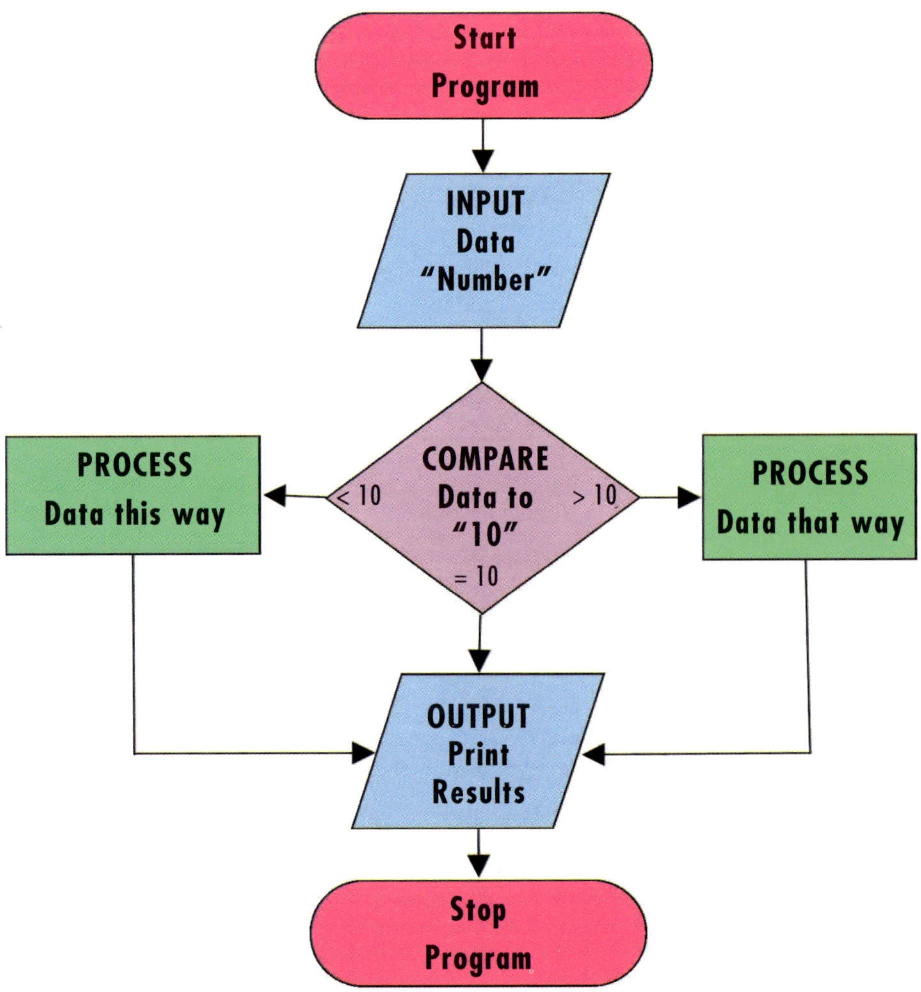

Walk the Walk Flowchart

Sample program that can be acted out using a room with floor tile or a pattern like graph paper.

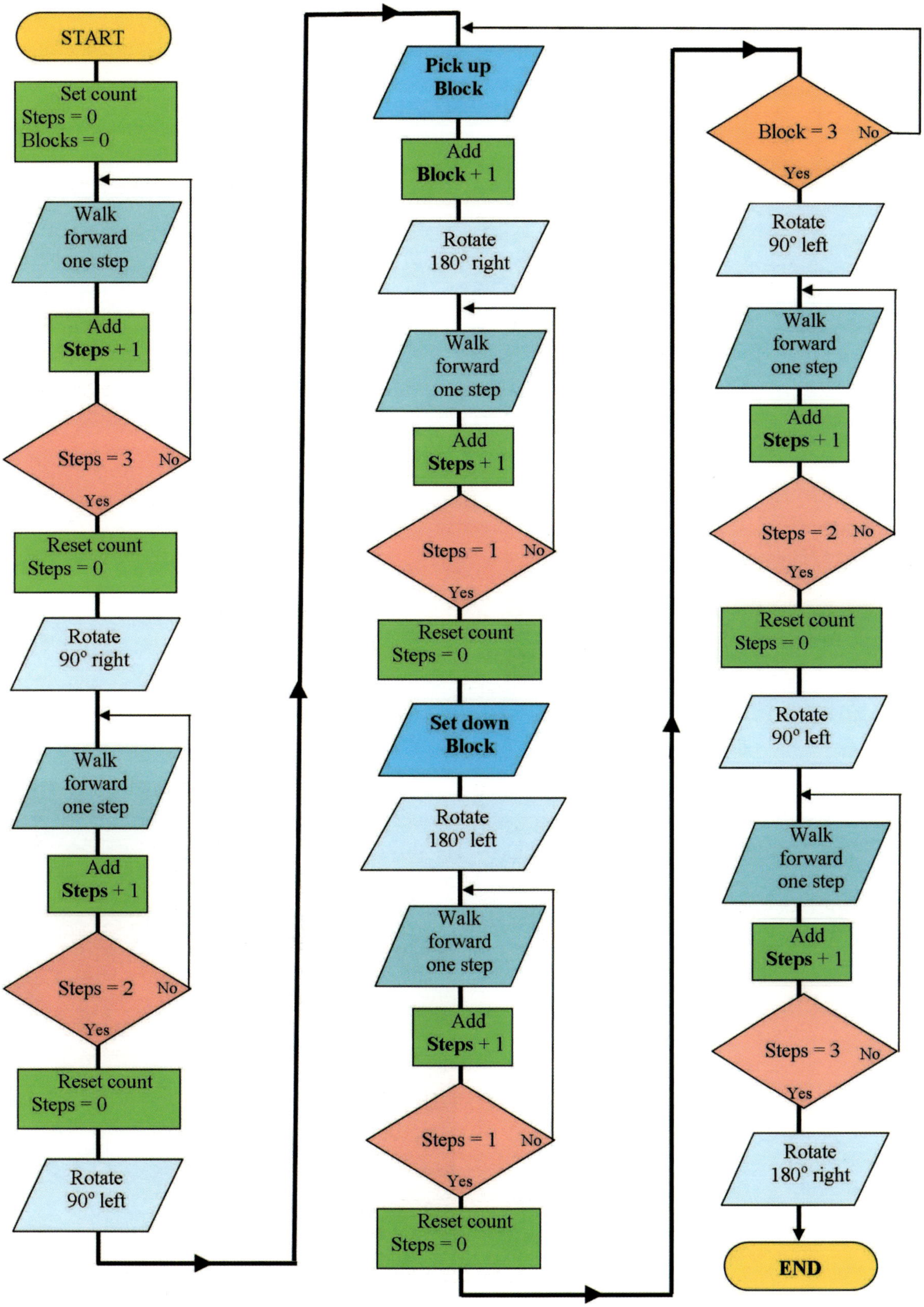

Walk the Walk Floor Grid Setup

Use floor tile or other markings to create a grid.

			Pickup Blocks	
			Pickup Blocks	
	↑ Actuator Start/End			
	Processor	Data Memory steps	Data Memory blocks	

Position locations for the four team players — actuator, processor, data memory for steps, and data memory for blocks — are shown. Also designate a position for a chair and/or a box for blocks or balls.

Walk the Walk Instructions

The Walk the Walk Flowchart

Using the Walk the Walk flowchart, have teams of three or four act out the program.

Actuator – One person will be the *Actuator* or movement of the robot and will stand on the start/end box facing away from the *Processor.*

Processor – The second person will be the *Processor,* who will read the flowchart program and give commands to the *Actuator* and *Data Memory.*

Data Memory (1 or 2) – The third and/or fourth person will be *Data Memory* to store variable data input/output. If the team has only three members the *Data Memory* will store both Steps and Block data. If the team has four members, one person will store Steps data and the other will store Block data.

How to Play Walk the Walk

The Processor will start by reading the Walk the Walk Flowchart program one box at a time.

Example:

- *Processor* to say out loud to everyone: "**Start Program.**"
- *Processor* to say out loud to *Data Memory:* **"Set count; Steps equal zero; Blocks equal zero."**
- *Data Memory* will write count on Memory Data Cards.
- *Processor* to say out loud to *Actuator* the next command: **"Take one step forward."**
- When complete, the *Actuator* will say out loud to *Data Memory:* **"Add one to the current Steps count."**
- *Data Memory* will write new count on Memory Data Card.
- *Processor* will ask *Data Memory* out loud: **"What is the step count?"** *Data Memory* will check the data sheet; add up the step count, and say out loud the **"number"** to *Processor.*
- *Processor* will check to see if it equals three.
  - **if it equals three,** then *Processor* moves on to the next program command step.
  - **if it does not equal three,** the *Processor* will follow the arrow back in the program and repeat the box to ask *Actuator* to take another step.

Process continues until reaching END command box.

Walk the Walk Memory Cards

Data Memory: Blocks

	Value	Value	Value	Value	Value
Count 1					
Reset 2					
Reset 3					
Reset 4					
Reset 5					

When told by Processor, Data Memoory will track current "Count" value for the variables.

⇨ Record new value and cross out old value.

⇨ Use new line when counter is reset to zero.

Count Example	~~Ø~~	~~1~~	~~2~~	~~3~~	
Reset Example	~~Ø~~	1			

Data Memory: Steps

	Value	Value	Value	Value	Value
Count 1					
Reset 2					
Reset 3					
Reset 4					
Reset 5					
Reset 6					
Reset 7					
Reset 8					
Reset 9					
Reset 10					
Reset 11					
Reset 12					

Activity R – Say What? Design Team

Performance Task For Youth

You will plan and design a program using the flowchart format involving basic commands like loops, goto, and if-then-else so that it can be acted out on a grid.

Success Indicator

Youth will create a flowchart program that includes standard symbols and actions to act out if-then-else, goto, and loops.

List of Materials Needed

- Robotics Notebook
- Activity Supplies
  - Same items from Activity Q, Walk the Walk
  - Copies of the handout/poster of the Flowchart Elements on page 94
  - Blank layout grid paper if not using the Robotics Notebook

Activity Timeline and Getting Ready

- Activity will take approximately 20 minutes.
- Use the same teams from Activity Q, Walk the Walk.
- Obtain all necessary materials. Participants should have paper to write on.

Experiencing

1. Lead a group discussion on how to use the flowchart shapes to make a program. Groups may want to use cutouts of the flowchart element shapes and create a flowchart of the steps. They also may want to use a grid to indicate travel and movement, and work backward to the flowchart.
2. Use the Walk the Walk grid as the work area, and the flowchart as a starting point. Have each team modify the Walk the Walk flowchart to perform an action or movement different from the one they did in Activity Q, Walk the Walk.
   a. Have the teams record their ideas and make a list of instructions (a flowchart) in their Robotics Notebook.
   b. Once the team has a flowchart and a grid setup (locations of chairs, start/end, etc.), they should make the grid and test (debug) their flowchart.
   c. Based on the test, they should make changes or modify the flowchart, if necessary.
   d. On a separate piece of paper, make a copy of the flowchart program.
   e. Have the teams exchange their flowchart program with a different team.

Sharing and Processing

As the facilitator, help guide youth as they question, share, and compare their observations. Have participants record their experience in their Robotics Notebook, including notes about new discoveries and knowledge. If necessary, use more targeted questions to prompt particular points.

- Ask the teams to share the types of modifications they made in their flowchart.
- What other commands might make it easier to program their robot?

Generalizing and Applying

- Have the teams compare their instructions with those provided by another team. What are the similarities and differences?
- Youth can apply what they have designed in Activity S.

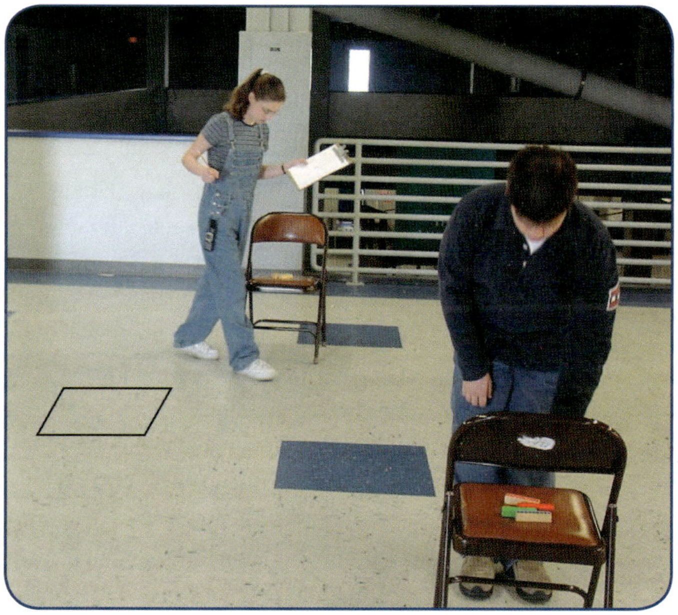

Activity S – Say What? Build Team

Performance Task For Youth

You will follow the program designed by another group and offer suggestions for improvement.

Success Indicator

Youth will be able to review, follow, and make improvements to another team's flowchart program.

List of Materials Needed

- Robotics Notebook
- Activity Supplies
  - Copy of a flowchart from Activity R that was designed by another team
  - Area with grid for acting out the program (same as in Activity R)
  - The same materials, chairs, and handouts used in Activity R

Activity Timeline and Getting Ready

- Activity will take approximately 20 minutes.
- Use the same teams from Activity R, Say What? Design Team.

Experiencing

1. Have the teams use the flowchart and grid developed by one of the other teams.
2. Have the team act out the new program and verify that it is accurate and that it works.
3. The team should debug the program, if needed, and make suggestions to improve or change the program.
4. The teams should return the programs to the original team for review of comments.
5. The teams should record the suggested modifications and actions in their Robotics Notebooks.

Sharing and Processing

As the facilitator, help guide youth as they question, share, and compare their observations. Have participants record their experience in their Robotics Notebook, including notes about new discoveries and knowledge. If necessary, use more targeted questions to prompt particular points.

- Share the basics of a flowchart or program.
- Have the teams share their experience in running their programs and running another group's program.
- Why do you think another team might find errors even if it had already been debugged?
- How could you change the program to make it better, more complete, or clearer to follow?

Generalizing and Applying

- How was the program following the floor graph similar to using the XYZ coordinates in the Level 1 activity?
- How could these flowcharts be made more accurate or precise?
- Many computer languages allow for comments between the commands. How could comments have been helpful in this activity?
- Additional challenges:
  - Have participants design a flowchart to make a peanut butter and jelly sandwich, a paper airplane, or another simple activity.
  - Use a computer to write a simple program for a robot, using RoboLab, BASIC, or another programming language.

Module 5: Ready, SET, Go!

Overview of Activities in this Module

To Learn
Activity T – Build Your Robot Design Team

To Make
Activity U – Build Your Robot Build Team

Note to Leader

In this module, youth will design and create their own robot using all the knowledge they acquired from past activities. Now that the youth have explored the world of robotics, they are ready to use what they have learned. Here is your chance to help them plan and design their own robot. Youth are free to construct their robot any way they like. Help the youth figure out which tasks they would like their robot to perform before entering the design phase. Do the youth want a robot vehicle? A robot to explore the land or underwater environment? A robot to sense its surroundings and respond in some manner?

Many books and materials provide additional ideas and suggestions. You may want to encourage youth to use building sets such as K'nex® or Lego®. Track 3 of this curriculum can help youth explore robots using these kits. Homemade robots also can be created in various ways using hardware store supplies or household materials.

Have fun, be safe, and help youth build their robot!

What You Will Need for Module 5: Ready, SET, Go!

- Robotics Notebook for each youth
- Activity Supplies
  - Trunk of Junk, see page 8
  - Supplies will vary and will depend on the type of robot the youth decide to build. Have an assortment of supplies from previous modules available.
- Toolbox

Timeline for Module 5: Ready, SET, Go!

Activity T – Build Your Robot Design Team

- Activity will take approximately 60 minutes.
- Divide youth into groups of two or three.

Activity U – Build Your Robot Build Team

- Activity will take approximately 60 minutes.
- Use the same groups as those in Activity T, Build Your Robot Design Team.

Focus for Module 5: Ready, SET, Go!

Big Ideas

- Robots can sense, plan, and act. They generally are designed to perform a specific job better than it could otherwise be done by a person or machine.
- How robots behave or act is a product of their design. The robots' structure, components, and programming determine their functions.
- Engineering design is a purposeful process of generating and evaluating ideas that seek to develop and implement the best possible solution to a given problem.

NSE Standards

- Abilities necessary to do scientific inquiry
- Evidence, models, and explanation
- Abilities of technological design
- Understanding about science and technology

STL

- Engineering design
- Problem solving
- Apply the design process

SET Abilities

- State a Problem
- Invent/Implement Solutions
- Redesign

Life Skills

- Contribution to group effort
- Problem solving
- Keeping records

Performance Tasks For Youth

You will design an original robot that can perform a specific task.

You will take your robot design and build a working model.

Success Indicators

Youth will have a design with details on how to build their robot, including steps, parts, and function.

Youth will have an original, working robot that will perform a specific task.

Activity T – Build Your Robot Design Team

Performance Task For Youth

You will design an original robot that can perform a specific task.

Success Indicators

Youth will have a design with details on how to build a robot, including steps, parts, and function.

List of Materials Needed

- Robotics Notebook
- Activity Supplies
  - Trunk of Junk
  - Activity supplies will vary, depending on the type of robot the youth decide to build. Have an assortment of supplies from previous modules available.

Activity Timeline and Getting Ready

- Activity will take approximately 60 minutes.
- Divide youth into groups of two or three.

Experiencing

1. Discuss with the group the different topics that were covered in the past activities; include the robotic arm, friction, electricity, and programming.
2. Pose this question and allow the groups to discuss and decide: What task would you like your robot to perform? Have the groups share their task and explain why a robot would be better at performing this task.
3. Youth will design an original robot that will perform the task.
4. Youth will then present their design to the group, explain what the robot will do, and why they went with their design.

Sharing and Processing

As the facilitator, help guide youth as they question, share, and compare their observations. Have participants record their experience in their Robotics Notebook, including notes about new discoveries and knowledge. If necessary, use more targeted questions to prompt particular points.

- Were there any designs you thought of and then discarded? Why?
- Are there daily tasks for which a robot would be helpful?
- What are some things that can be made easier by using a robot?

Generalizing and Applying

- Could this robot be helpful in the real world? Is there a need for this robot?
- Youth can apply what they have learned in the next activity.

Activity U – Build Your Robot Build Team

Performance Task For Youth

You will take your robot design and build a working model.

Success Indicator

Youth will have an original working robot that will perform a specific task.

List of Materials Needed

- Robotics Notebook
- Activity Supplies
  - Trunk of Junk
  - Activity supplies will vary, depending on the type of robot the youth decide to build. Have an assortment of supplies from previous modules available.
  - Depending on the designs that the youth develop, they may have to help collect the materials they need.
- Toolbox

Activity Timeline and Getting Ready

- Activity will take approximately 60 minutes.
- Use the same groups from Activity T, Build Your Robot Design Team.

Experiencing

1. Have youth build their robots. Make sure they use information they have learned in the past activities.
2. Have youth record in their Robotics Notebook any changes they made to their design — what worked and what didn't.
3. Allow groups to test their robot and redesign, if necessary.
4. Have youth give a presentation about their robot, including what it does, why they came up with this model, and future improvements they would like to make.

Sharing and Processing

As the facilitator, help guide youth as they question, share, and compare their observations. Have participants record their experience in their Robotics Notebook, including notes about new discoveries and knowledge. If necessary, use more targeted questions to prompt particular points.

- How does your robot differ from your original design? Why did you have to modify your plans?
- How was building your robot harder or easier than you originally thought?
- What might you do differently next time?

Generalizing and Applying

- How might you use your robot in the real world?
- How could you design a different robot to complete the same task?
- What else could you use your robot for?

Glossary

Module 1

- **Battery** – a device that can store electrical energy that can be used at a later time
- **Electric circuit** – the complete path of an electric current, typically consisting of a power source (battery), a load, and an on/off switch
- **Electronic circuit** – a circuit that includes active control components like transistors that can switch electron flow automatically based on programmed logic
- **Parallel circuit** – a circuit containing multiple paths for the electricity to flow
- **Series circuit** – one continuous loop from a battery through the load and back to the battery
- **Switch** – a device for opening or closing a circuit
- **Double pole double throw switch** – a switch that can control two circuits in two positions; can be easily wired to reverse the electrical flow
- **Double pole single throw switch** – a switch that operates two circuits at the same time
- **Single pole double throw switch** – a switch that controls two circuits, one at a time
- **Single pole single throw switch** – simplest switch that opens and closes one circuit

Module 2

- **Accuracy** – in engineering, the closeness to the actual or true value of a measurement or location of movement
- **Analog** – a signal that is continuous in nature with many variations as in temperature, wind, or pressure; might be thought of as a playground slide where readings could be taken at any point along the slide with almost unlimited possibilities
- **Digital** – a signal that has specific levels or steps; might be thought of as a set of stairs with readings that can only be taken at the steps, 1, 2, 3, 4, 5, etc.
- **Flowchart** – a diagram that represents a process, showing steps as boxes and their order with connected arrows
- **Precision** – in engineering the repeatability to produce the same results in multiple similar actions
- **Sensor** – a device that measures a physical quantity, such as heat, light, motion, or barriers, and converts that quantity into a signal that can be used by an instrument
- **Sensitivity** – being able to detect and respond to small amounts of information

Module 3

- **Algebra** – a type of mathematics that uses a set of rules for operations and relations in solving problems to find unknowns or variables
- **Binary** – a number system represented by only two symbols, 0 and 1
- **Boolean logic** – name for the logical system (algebra) used in electronics in which inputs are evaluated by logical operators
- **Breadboard** – a circuit prototyping device that allows components to be connected for testing without soldering
- **Capacitor** – an electrical component that holds an accumulated charge of electricity
- **Diode** – an electrical component allowing current to flow in only one direction
- **LED** – Light Emitting Diode; a diode that will produce light with very little energy
- **Logical operator** – a function, such as OR, NOT, AND, used to evaluate two binary inputs and issue a binary output
- **Resistor** – an electrical component that restricts the flow of electrons in a circuit, thus decreasing voltage
- **Transistor** – an active electrical component used in a circuit as an amplifier, detector, or switch

Module 4

- **Actuator** – a part used for motion control or movement; a robot arm, or gripper, or the power used to move a piece of the robot
- **Data memory** – an electronic storage area for information to be written, used, changed, or reused; processor helps manage where and when the data is stored and used
- **Debugging** – testing and modifying program code to ensure the desired result
- **Processor** – also called a microprocessor or Central Processing Unit (CPU); the part of the computer that reads the program instructions, makes decisions, and provides feedback; works with other parts of the system such as memory, input, and output; used in many appliances, cars, phones, toys, and more because the CPUs have become so small
- **Program** – a sequence of instructions that perform a specific task for an electronic device